THE NEWSPAPER THAT LINES THE BOTTOM OF A BIRD CAGE AND OTHER STORIES FROM THE EMERGENCY DEPARTMENT

EILLYNE SEOW

Illustrators

Goh Hsin Kai
Alicia Vasu

This book is dedicated to our fellow warriors in emergency departments all over the world, and to the people whom we have the privilege to care for.

CONTENTS

Acknowledgements

Many thanks to our friends who shared their stories:-
Wui Ling, Swee Hui, Seow Siang, Mic Chia, Charmaine M, Margaret N,
Chui Mun, Honey, Tsuan Hao, Wei Kian, Dennis, Eu Jin, May Ling,
Eva, Jo Keow, Hsin Kai, Wilson C, Kay Fei, Yew Kan, Daniel K, Albert Y,
Kim Chai, Vivian S, Shu Woan, Wee Yee, Gregory C, Chee Kheong.

Special thanks to Honey, Wen, Fuh Yong, Alicia, Poi Leng, Jenny Q, Yew Nah
and Grace for their assistance in the production of this book.

Thanks also to Greg Henry for providing the title of the book.

Along the Corridor

Over the years I have been asked several questions.

"What do you work as?" Doctor.

"Where do you work?" A hospital.

"What department?" The emergency department.

There are different reactions to my answers: blank looks, disappointed looks, nods of acknowledgement, and questions such as "Are you a specialist?" and "Is it (emergency) a department?"

Initially, I understood the reaction. Emergency departments were regarded only as entrances to hospitals. Later, with the TV series *ER* some regarded it as a glamorous place of action. However, even our own medical community took a while to realize that the teams in emergency departments have moved on from being sorters and postmen to sievers (using 'sieves' with different aperture sizes), gatekeepers, resuscitators, diagnosticians, counsellors.

The emergency department (ED) is a high-stress, high-octane environment where the tempo changes without warning. An old friend used to describe her shifts as "going through changing seasons in eight hours". To me 'unpredictable weather' was a more apt description after all seasons follow a predictable sequence – spring, summer, autumn, winter – unlike work in the ED.

We decided to write this book because even today, the community at large and our own medical community have only a vague idea of the encounters that occur within an emergency department. We thought about how we should introduce the people and the place, share with you the flavours, the smells, and the emotions that churn within. We used Hilary Mantel's method in constructing this book "like a collage: a bit of dialogue here, a scrap of narrative, an isolated description".

This book can be read in sequential chapters but feel free to dip in and out of the various stories. We hope that even after a few pages, you will begin to feel the pulse of the place we work in.

The encounters have been fictionalized and any resemblance to any persons is the reader's own conjecture.

MY STORY

"If you like her, you will discourage her from doing emergency medicine," a young consultant said to me on one of my shifts.

"Her" was a young medical officer who, in my judgement, would do well as an emergency physician. She was able to think on her feet, could multi-task, and could draw relevant information from various sources to arrive at a logical conclusion.

The young consultant was one of the best emergency physicians in her cohort, but she was not a happy emergency physician.

What is the disadvantage of emergency medicine compared to other disciplines? Emergency physicians cannot be their own bosses. Sure, they could open a general practice and most would be pretty good at it. Emergency physicians worth their salt are above-average diagnosticians and are the best detectives in the pack, like Sherlock Holmes or Detective Dee. However, if they want to practice emergency medicine, they have to work in an institution, whether it is public or private. This means that there are bosses who are the best – ones who run systems that allow their emergency physicians to see patients with little or no effort. This also means there are bosses who are the worst – ones who run systems that are illogical and inconsistent, making it difficult for emergency physicians to deliver safe patient care in an effortless manner.

"I did emergency medicine because I want to be a real doctor," one of my colleagues shared over drinks during one of our regular get-togethers.

He had a master's in sports medicine and could have gone to a lucrative private practice.

It is true that, you have to be trained in and see enough emergencies to be able to respond adequately when faced with one. Any emergency physician at the end of his training is able to respond to an emergency and manage any resuscitation; it is part of our DNA. Those with the right temperament require very little training to get into the groove of things, most gain it by sheer exposure, and a few have been shell-shocked.

A young doctor who was on the dean's list in his university years took sometime to get into the groove. It was a struggle, especially when he had to manage a few resuscitations simultaneously, but our young doctor finally made it.

"So why did you do emergency medicine?" I asked another young colleague.

"Emergency physicians are altruistic," she replied. "I told my surgical friends that we are more altruistic than them."

I laughed. I do not agree that we are more altruistic than our surgical colleagues who work very hard in the public institutions.

What motivates doctors to train as emergency physicians? I have heard more than one emergency physician moan, "I should not have done emergency medicine." Another said, "In my life, there are two decisions I regret making. One of them is specializing in emergency medicine."

Who in their right minds would do emergency medicine? After all, doctors are some of the cleverest people.

An emergency physician always has a boss. He will earn a comfortable living but will never be as rich as his oncologist or surgeon friend. He most likely will be doing shift work till he retires, unless he decides to be a professional locum. He should forget about having the working hours of his dermatology or ophthalmology friends.

He will always have to keep abreast of the latest in medicine, covering the breadth and (some) depths on all subjects; he is a life-long learner. He will meet a wide spectrum of society most of the time at their worst, and at times their most stoic. He will have to comfort when he cannot heal. He will have to manage patients who present with non-specific complaints, teasing out information from them and their accompanying persons; he must decide which to believe and which to discount. He will have to save patients when he is not certain what is killing them.

By now you may be wondering whether emergency physicians are masochistic. You may be right. Why did most of us become emergency physicians, other than being conned? I admit that I did con quite a few of my juniors.

I entered medical school intending to open a general practice and be a family physician. I liked people so I thought being a family physician would allow me to know many more stories (being the kaypoh that I am), help patients get better, earn as much or as little as I wanted and be my own boss.

Then I did a posting in emergency medicine. I saw many types of people who most times were not at their best, but the relief on the faces of those I had treated was worth all the annoyances of that shift. Even the shenanigans of the boss could not take away the feeling of having made a difference.

Patients do not come in with a diagnosis of heart attack written on their foreheads. It is a challenge to arrive at the correct diagnosis especially when information is disparate, inconsistent or unavailable. However, when we arrive at the correct diagnosis, despite all the obstacles, that feeling of gratification mixed with triumph is priceless.

One of my former bosses visited a European country and was much taken with the system of their emergency departments. There was even a rumour that he wanted to change the present system to the one he saw. The system he saw was the old world one where a staff at the door would direct the patient to which department he thought could treat the patient's complaint.

The obvious complaints are easy like the patient falling on his outstretched right hand: there is a deformity of his right wrist, and he sees the orthopaedic specialist. Another patient complains of chest pain and is sweating: this patient sees the cardiologist because an acute myocardial infarct (heart attack) has to be excluded although sometimes the cause may be a reflux of acid from the stomach.

A patient complaining of stomach pain or epigastric pain may be channelled to the gastroenterologist or the internist but I have known of patients with acute myocardial infarcts presenting with only this complaint. It gets a bit more challenging at the door when an elderly person presents with giddiness. The internist or geriatrician will then have to figure out what the cause is.

The way emergency medicine was practiced in the old-world system was the best the system could offer before the conceptualisation, development, and establishment of the present system of emergency medicine. Back in the European country, whose emergency department system, my former boss was much taken with, that country moved towards developing and establishing emergency medicine the following year. The last I heard, a few of their emergency departments now had trained emergency physicians, similar to the emergency departments in Singapore, the United Kingdom, the United States, Hong Kong, and Malaysia.

What else stopped me from being my own boss? It was pulling a patient from the jaws of death. My heart sped as I raced to intubate (put a tube into her airway to help her breath) a young asthmatic girl before she stopped breathing. A fellow doctor gave the sedative followed by the paralytic agent. The team waited for the paralytic agent to take effect. It was nerve-racking as her oxygen saturation started to drop from 100 per cent towards 95 per cent. When we saw fasciculation of the muscles of her lower limbs, we moved into action. The endotracheal tube was in and fastened securely; the movements were synchronized and flew into each other. The reading on the oximeter showed oxygen saturation climbing back to 100 per cent. A few days later, this young girl walked out of the hospital, completely well.

There were more cases over the years that affirmed my decision to be an emergency physician. I attended to an eleven-year-old boy who was drowsy after being hit by a car. I made the decision to intubate him; he survived and also walked out of the hospital.

Another was a forty-year-old woman who came to the emergency department complaining of vomiting and diarrhoea. While waiting to be seen, she complained of being tired. A nurse brought her into the consultation area to let her lie on a trolley. It was 7.30 p.m., and I had finished my paperwork. I walked in to the consultation area to touch base with the doctors on duty. This forty-year-old woman threw a fit as she was being assisted onto a trolley, and the nurse accompanying her shouted for help. We ran towards the patient, and I put my finger on her neck to look for a pulse. There was no pulse. The patient was moved to the resuscitation area and defibrillated (electrical shock applied to the heart). We started chest compressions and continued in between the three shocks we gave her. The patient's heart had gone into ventricular fibrillation (the heart muscles were quivering in an uncoordinated manner and no blood could be pumped out of the heart). At the end of three shocks, the patient gave a gasp and her heart returned to normal. We could feel her pulse in the neck. This patient walked out of the hospital well.

I was not successful in all the resuscitation cases I managed. A few left me frustrated and in despair. I lost a thirty-year-old woman with low blood pressure. I did not know what her problem was and was unable to bring up her blood pressure. Another was a patient who walked into the emergency department and then collapsed - no pulse, no breathing - while waiting for the results of his investigations. No matter how hard and long we tried, we lost him. A six-month-old was brought in with no pulse and was not breathing: despite heroic efforts, we were unable to snatch him back. We hid the despair we felt from the young, stoic parents. We spoke quietly and held back our tears for their sakes. There were other occasions when I had to hold back my tears because it would be unseemly for me to be seen crying on the shop floor.

Patients and their relatives are rarely at their best in the emergency department. A few may be at their most stoic and there had been a few heroic moments. On one of these occasions, the young consultant whom I mentioned at the beginning of my story, nearly cried openly on the shop floor. A young girl who had lost her right hand in an accident requested this young consultant to tell her mother not to cry or feel sad, and to reassure her mother that she was all right.

Do I have any regrets about being an emergency physician instead of my own boss? No, I have none.

Would I encourage young doctors to be emergency physicians? Should I? I posed this question to Mic, one of those I had conned into being one. He had been deciding between emergency medicine and radiotherapy. We had supper at Swee Choon Tim, a dim sum supper place along Jalan Besar. Those days the restaurant occupied only two shophouses. It was full inside and we had a table along the back alley next to the drain. We spoke about his likes and dislikes, his strengths and weaknesses, and his suitability for each specialty.

"I would present them with the pros and cons of emergency medicine practice and let them decide," Mic said. Did he have any regrets about choosing emergency medicine? Mic had none.

Sandra, the young medical officer I spoke about earlier, has invited me for dinner. She is now at a crossroads in her career. One fork leads to a family medicine practice and the other heads to emergency medicine.

Do I like her? Yes, I do.

Part 1

THE PLACE

CHAPTER 1

THE NEWSPAPER THAT LINES THE BOTTOM OF A BIRD CAGE

One of my gurus told me that the emergency department is the newspaper that lines the bottom of a bird cage. In other words, we catch a lot of shit.

"Who the hell do you think you are? I know Minister So-and-so. Fuck you!" It was a typical weekend shift, but this gentleman seemed to have arrived early. He stank of alcohol and vomitus. The two police officers who had escorted him to the emergency department kept their faces devoid of expressions as they stood at the side of the trolley.

This gentleman was on a trolley in a corner of the trolley area. Unfortunately, it was an open space chock-a-block with trolleys, although the staff did leave a visible space between this gentleman's trolley and his nearest neighbours. The relatives of other patients who were milling around had stopped in their tracks to stare at the gentleman, agog at the commotion he was causing.

The gentleman continued with his colourful expletives, flailing his limbs around, tossing about on his trolley and declaring his friendship with a certain minister. Fortunately, we had a strapping health assistant, Rodney, on duty that shift. He was tasked to change and clean up the gentleman. Fortuitously, the patient zonked out and became a dead weight just before he was to be cleaned up.

While he was "cooperative", we swiftly got intravenous access, started fluids, obtained a random blood sugar, and performed a head-to-toe examination.

"He's got a 'baluku', a big one, in the occipital region – five centimetres," I announced as I felt around the gentleman's scalp. When I lifted my gloved hands from his scalp, they were bloody.

Eric, the nurse was about to set a line on this gentleman but he paused, holding the intravenous cannula mid-air. He raised his eyebrows and asked, "T&S (toilet and suture) or staples? CT?"

I nodded and muttered, "Staples faster, I think." I continued with the rest of the gentleman's physical examination. There was no other obvious injury found. "Okay, nothing remarkable. Let's staple the wound and send him for a CT head quickly. I hope he doesn't wake up before it's done."

The two police officers must have been experienced guys. When they heard CT, they looked at each other and nodded. One went to make a phone call. It was close to the end of their shift.

The gentleman was duly sent off for a CT scan head, which did not show any abnormality. We decided to admit him to the short-stay ward for monitoring and also, for him to sleep off the effects of the alcohol; this was known as "lodge till sober".

A few hours later, this gentleman who had been sleeping like a lamb, started tossing in his trolley. When we tried to calm him, he started to flail his limbs again and the expletives came fast and furious.

"Your drunk gentleman is turning aggressive," Lili, the young nurse from the observation area informed me breathlessly. She looked flustered, and her hair, which had started at the beginning of her shift in a neat chignon, was now in a ponytail.

I was surprised to hear that he was still in the emergency department. I had expected him to have reached the short-stay ward by then. "Vitals?"

I asked Lili as I entered the observation area and walked towards the gentleman. He was still tossing about in his trolley.

"Stable," Lili answered.

"BP? Pulse rate? Temperature? Oxygen saturation?"

"It's 140/70, 97 per minute. Sorry I didn't take the temperature or oxygen saturation," Lili answered sheepishly. "Will do it now."

I looked at her with a raised right eyebrow.

"Let's see if I missed anything during my examination," I said softly as I started to prod gently at the gentleman's head, neck, chest and abdomen. "Ah, so *this* is the reason," I muttered as I felt a lump over his lower abdomen – to be precise, his hypogastrium. It was then I noticed the number five written on the saline bag hung on his trolley. "Is this the fifth pint of saline?" I asked flipping the normal saline bag.

Lili nodded.

"Rodney still around? Eric?"

The gentleman was now tossing even more restlessly and touching his crotch.

"Rodney, you are going to help get him to the urinal," I said.

After five minutes, while I was typing my notes, Rodney approached us at the counter. "He is not passing," Rodney said in bewilderment.

I looked up at him and asked "Did you sit him up and say shhh, shhh, shhhh?"

Rodney shook his head.

"Eric, can you give him a hand?" I asked.

They pushed the gentleman's trolley towards the toilet, screened him off from the rest of the patients, sat him up and started to whistle.

"Mission accomplished," Eric declared triumphantly.

The gentleman remained a gentleman for the rest of his stay in the emergency department, and he went home well. We were lucky.

* * *

Mr. V never behaved like a gentleman - in fact his behaviour sometimes bordered on that of a hooligan. He visited us on our night shifts and rarely during daytime. He had been released from prison a year ago after having spent several years inside. He had been a regular patient of ours even before his prison term. He'd had poor health ever since we knew him. He continued with his habit of smoking two packs of cigarettes a day while in prison despite being repeatedly advised against doing so.

Unfortunately after his release from prison, he developed an ulcer over his left foot which failed to heal. This was likely due to his badly controlled diabetes and poor peripheral circulation from his smoking habit.

He visited us once every few weeks, and sometimes more frequently. Our new doctors would know him very early in their postings because he was infamous for arriving and demanding to be attended to first. He would shout and make a ruckus, and when it was his turn, he was nowhere to be found. He would turn up a few hours later smelling of cigarettes and alcohol.

The team was concerned about infection, which this patient invariably had with his non-healing foot ulcer. An intravenous cannula would be inserted for antibiotics to be delivered. However, Mr V had the habit of disappearing again after this, and he would sometimes return after another few hours, or sometimes not return until the next day. His absconding would result in more paperwork for the team because his disappearance had to be reported to the police, especially when he left with an intravenous cannula still in-situ.

Patients like Mr V challenge our caring and professional instincts to their limits. They are like the shepherd boy who cried, "The wolf is here!" We constantly guard against this bias and warn our juniors to do so as well. It is hard especially when patients are verbally abusive, are physically threatening, smell of alcohol, and generally are not interested in their own welfare.

One general practitioner must have been shocked by my behaviour when he accompanied one of our regulars in an emergency ambulance. Those were the days before our emergency ambulance colleagues gave out the identification number of the critically ill patients they were conveying to us.

"Hotel four, hotel four, standby to receive male, Indian, with chest pain."

This was the message we received from the emergency dispatch centre one afternoon. As was our practice, two doctors and two nurses prepared to receive this "critically ill" patient. One nurse and one doctor (me) went out to meet the ambulance team when we heard the sirens. The general practitioner and the paramedic jumped out of the back of the ambulance first. I took a look at the patient as they unloaded him. The patient was slumped on the trolley, and before the general practitioner or the paramedic could say a word, I removed my gloves and turned my back on the patient. I walked in to the resuscitation area, threw my gloves into the bin, removed the apron I was wearing, threw that into the bin as well, and said to the other doctor waiting to receive the patient, "It's Mr S. Carry on. Any problem, call me."

Mr S was a familiar face whose chief complaint was chest pain. His portrayal of a patient with a heart attack was superb, but he lacked a good make-up artist, which was what gave away his acting. I would have nominated him for an Oscar if I knew how.

* * *

The hours to midnight on a New Year's Eve are generally quiet ones in emergency departments, in most parts of the world. However, a quarter of a century ago, in a particular emergency department, the crowd arrived much earlier than expected. It was also unusual as most were drunk. *(The inebriated patients generally arrived after ringing in the New Year.)* The team worked at sorting and treating the patients. They soon ran out of trolleys and had to borrow more from other departments.

The team on the afternoon shift worked over time and did the countdown to the new year with the night team.

Mr John S was brought in by the emergency ambulance team just after the countdown. He had been drinking and had vomited and dirtied himself. The emergency ambulance team handed over a bag containing Mr John S's medicines to Nurse Amy who received Mr John S. It was noted that there was insulin in the bag which alerted Nurse Amy that Mr John S was a diabetic. The bag was put into the cupboard, where patients' belongings were kept for safekeeping. A blood sugar level was obtained for Mr John S; this was normal. Another nurse washed and cleaned him up, and he was put onto a trolley to sleep off the alcohol.

At 4.00 a.m., the emergency department stank of alcohol. There was also a sour smell where patients had vomited despite the housekeeping staff vigorously cleaning these areas. Nurse Amy was busy but remembered that she had to check Mr John S's blood sugar level again. She asked her colleague Nurse Evelyn, to do it for her. Nurse Evelyn looked at the name tag around the wrist of Mr John S and checked his blood sugar. She reported to Nurse Amy that it was normal.

At 6.00 a.m., a few of the seemingly inebriated patients began to stir. The team continued with their scheduled two-hourly examination of these "sleeping" patients. By the time, the morning team arrived to take over the shift, most of the patients had woken up and were having hot tea and a piece of toast.

One of the nurses on morning duty approached one of the few patients still sleeping. She raised the alarm when she saw how blue he looked and how stiff he felt. The patient, Mr John S was moved quickly into the resuscitation area.

"Could you collect Mr John S's bag and pass it to the doctor in the resuscitation area, dear?" the nurse in-charge asked Nurse Evelyn.

"Where is Mr John S's bag?" Nurse Evelyn asked Nurse Amy.

"Why does the doctor need my bag, dearie?" a patient coming out of the toilet asked.

It then dawned on Nurse Amy and Evelyn that there were two patients with the same name, John S. They were stunned.

Insulin is a medication used to lower blood sugar. An overdose of insulin can cause a patient to have a low blood sugar level. When the sugar level is too low, the patient will become comatose and can die. A patient with very low blood sugar can look like a drunk.

* * *

The emergency department is open 24 hours a day, 365 days a year. It is where the police and the emergency ambulance teams brings the 'unknowns' and the dislocated.

It is the newspaper that lines the bottom of a bird cage.

* * *

CHAPTER 2

AN ANCHOR

Most of our regular patients will visit us after the sun has set. If they are hungry, they will arrive before we have finished serving all our bowls of porridge.

One of them, Mr Samy, would come in complaining of breathlessness. He had asthma exacerbated by his years of smoking. He would ask for his bowl of porridge and cup of milo. Afterwards, he would request for a bun, which we would give him. He would pocket the bun, smile, and thank all the doctors and nurses he met as he left the department with no fuss.

We would hear Ms. Chang's arrival. She would be singing a Mandarin ditty – off-key most times. Initially, I thought she was a man. She looked like a monk: head shaven bald, a yellow garb, and a necklace of extra-large beads. She would be in the ED for various reasons, but fortunately nothing serious. Ms. Chang would have her porridge and sing her ditties while waiting to be seen by a doctor. She would not harass the staff. When the ED became quiet, we would know that she had been attended to. She would say *thank you* and wave goodbye to all and sundry, including other patients and their relatives.

A few members of the team had bumped into her outside the Guan Yin Temple along Waterloo Street. She would greet them loudly as "Doctor" or "Nurse". They would sheepishly say hello to her and slide away. Ms.

Chang would wave and boom out, "Bye-bye, Doctor! Bye-bye, Nurse!".
She would then continue peddling her wares.

Then there was Mr Lee SM. The older nurses told us that he used to
be well-off till his nephew cheated him of his money, this was more than
10 years ago. He had asthma and was on long-term steroids. It showed
in his physique because he had very thin legs, his body was round, he
had a hump at the back of his neck, his skin was thin, and his hair was
brittle. He had a pleasant personality, however. He would come to the
emergency department when he had an asthmatic attack. He would
receive treatment, lie on a trolley overnight, and leave the next morning
when he felt better.

Mr Lee SM was happiest when we agree to hospitalize him. When
the hospital was short of beds, we would inform him and ask him to wait
patiently. At times, he would do a tour of the wards and return to report
to us which wards had empty beds. When patients had been waiting
for some time for a hospital bed, we would offer him a bed in another
hospital, and he would petulantly agree to go declaring that the food
there was better.

In the last few years, he would give out greeting cards during
Christmas and Chinese New Year to doctors and nurses he recognized. He
was also an excellent case to teach medical students with his florid signs
of Cushing's syndrome. He was an accommodating subject who would
patiently allow medical student after medical student to examine him.

However, Mr Lee SM had a pet peeve. He had a great dislike for a
particular patient; Mr Ng AL. Mr Ng AL, unlike Mr Lee SM had an
obnoxious personality. He would come in breathless because of his lung
and heart disease. When he was unwell, he would allow the nurses to
obtain his vital signs and electrocardiogram, and they would send him
for X-rays but when he was feeling better he would refuse everything.
He would then scold, curse, and use four-letter words on the nurses.
(Somehow, he was never heard scolding doctors.) He would sleep the night
through when he felt better.

Mr Lee SM would always complain to us about Mr Ng AL whenever they happened to be in the emergency department together. He would spy on Mr Ng AL, follow him to the toilet, and watch all his movements. Once he reported to us that Mr Ng AL was behaving suspiciously in the toilet and smoking "some white powder". We checked but could not find any evidence.

In his last month, Mr Ng AL's condition deteriorated to the point where we had to manage him in the resuscitation area. He collapsed and had no breathing or pulse. He was intubated, and we started chest compressions. He had a heart attack, but we managed to bring him back. He was admitted to the coronary intensive care ward. He was extubated the next day, and on the third day he discharged himself against medical advice. He returned to the emergency department, two days later, breathless. He was re-admitted and discharged himself again the next day. This happened a few times till his last visit. This time he refused to be hospitalized. He walked out of the emergency department and five hundred metres later, he dropped dead. The public called for an emergency ambulance, Mr Ng AL was brought back and declared dead on arrival in the emergency department.

* * *

The emergency department is open 24 hours a day, 365 days a year. It is where our regulars come for comfort and care, where they are on familiar grounds. It is an anchor in many of our patients' lives.

Eillyne Seow

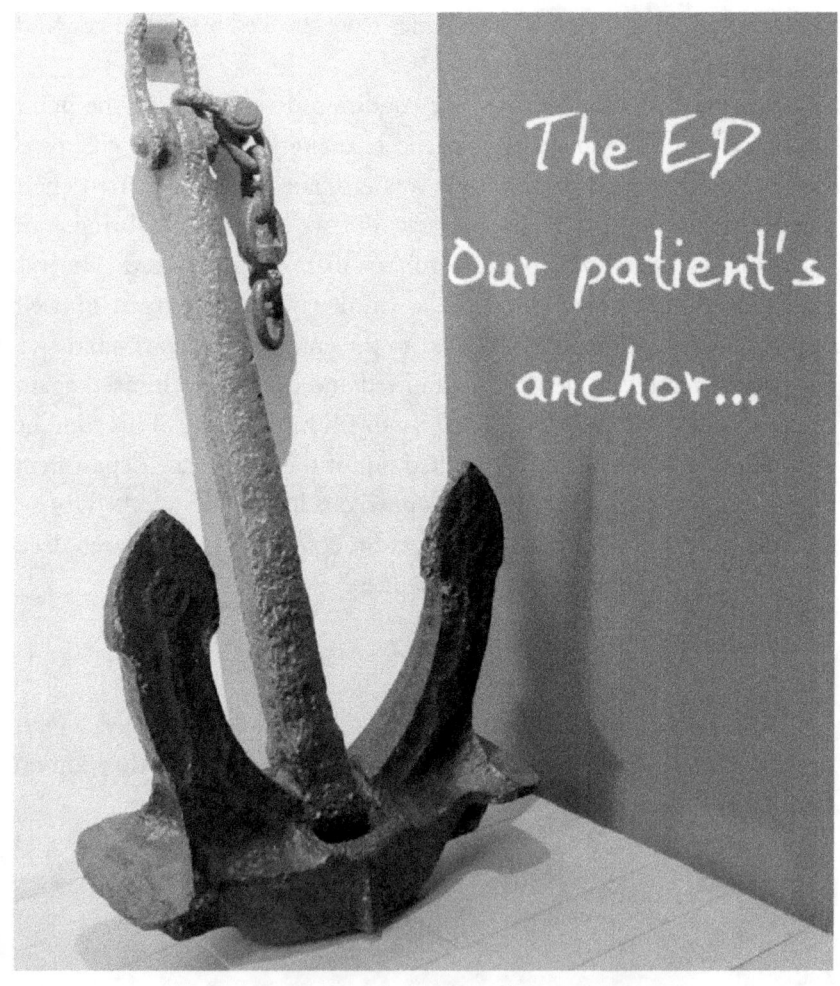

The ED
Our patient's
anchor...

CHAPTER 3

ED – A DUMPING GROUND?

Holidays are a joyous time for most but it is also when we see the worst side of families.

"Mr Soh is an eighty-year-old male, presented with ARU [acute retention of urine] three days ago, and had an IDC [indwelling catheter] inserted. He is back today. He pulled out his IDC and had an ARU again," my junior, Dr Tan, reported to me.

"Gentleman - eighty-year-old gentleman," I corrected as I rapped his knuckles.

"Oh! Sorry, eighty-year-old gentleman," Dr. Tan repeated with a straight face.

"The patient is not a dog - he is a gentleman," I remonstrated.

Dr Tan nodded, wisely keeping silent. It is not politically correct to ruffle your senior at the beginning of your shift.

"Has an IDC been re-inserted?" I asked.

"Yes," Dr Tan answered.

"Urine combur-9?" I prompted as I turned to look at Mr Soh's urine combur-9 results on the computer screen.

He stood behind me and pointed at the report showing on the screen. "Microscopic hematuria likely due to trauma, but otherwise nothing abnormal," he reported.

"Okay, what is your planned disposition?" I asked as I checked through his entries on the screen.

"Er...."

I stopped moving the mouse and swivelled to face him, giving him my full attention.

"Er...." he said as he cleared his throat.

There were two doctors and a nurse at the central counter with us. They all paused to look at Dr Tan.

I raised my eyebrows and inquired, "Mr. Soh can go home, right?"

"We are having some difficulty discharging him," Dr Tan said in a soft voice.

"Wait..." I said as I turned to look at the screen again. "Eh, this uncle was here six hours ago, so how come his case is only being flagged now?"

"That's the problem." Dr Tan sighed with relief and poured out the story.

Mr Soh had been conveyed by the emergency ambulance service at 10.00 a.m. that day for acute retention of urine. He was an eighty-year-old Chinese man who was lucid and knew his address. The ambulance crew had informed us that a relative of his had called 995 for assistance. After a urinary catheter had been re-inserted, Dr Tan had attempted to contact his relative but with no success. Although Mr Soh was fit for discharge, he could not go home unaccompanied. Dr Tan had contacted the medical social worker in charge of emergency department.

The medical social worker, Michael, had brought Mr Soh home but Mr Soh had no keys. There was a twenty-year-old boy inside the house who refused to open the door. The boy told Michael that his parents had instructed him to have his grandfather hospitalized till they returned from their holidays because there was no one to take care of Mr Soh. Michael had no choice but to bring Mr Soh back to the emergency department, and he requested that we admit the patient.

"Admit the patient?" I asked in an incredulous tone.

There was silence again at the central counter.

"Michael said that it's 5.00 p.m., and he is unable to find placement for Mr Soh because it is after office hours. He asked us to admit the patient," Dr Tan said.

I sighed, swivelled back to the screen and brought up the bed waiting queue. "We have ninety patients waiting for a bed. The longest wait is twenty-five hours, and we have to admit Mr Soh." I shook my head and sighed again. "Okay, you can admit him," I said in a resigned tone.

"By the way, when are his children coming back?" I asked.

"According to the grandson, after Christmas, 26 December," Dr Tan said in a soft voice.

I paused, "Ten days more?"

Dr. Tan stopped in his tracks. "Still admit?" he asked.

I nodded, resigned.

The team managed to contact the medical director of a community hospital the next day. The patient was still in the emergency department the next morning - not a surprise. The medical director agreed to place the patient in his institution, which saved a bed in an acute hospital and released space in the emergency department.

CHAPTER 4

REFUGE FOR THE
SILENT VICTIMS

Attitudes towards domestic violence have changed over the last two decades. A 2014 newspaper reported that there had been an increase in the number of cases seeking help, applying for personal protection orders and approaching the police for assistance compared to the year before — a good sign. Some of us worry whether there is an increasing number of spouses being abused but some of us would like to think that victims no longer think that they are at fault.

"How did you get these?" I asked Ms Ti softly as I gently palpated her abdomen. She had multiple bruises of different ages over her chest and abdomen. She winced as I touched her right loin. I stopped. "Could you sit up for me?" I asked. I lifted her blouse, a Banana Republic piece that in those years could not be bought in Singapore. I barely managed to stop myself from cursing.

There was a reddish bruise over her right back and multiple abrasions over the rest of her back. "How did you get these?" I asked again.

Ms Ti was at the emergency department complaining of pain in the abdomen. I had not suspected that this was going to be anything but a straightforward case.

She had walked in, an elegant twenty-eight-year old lady. She spoke in a well-modulated voice and in grammatically correct English. She complained of pain in her epigastrium, that was worse when she was hungry, relieved by food but in the last few days, the pain had increased in frequency and severity.

"I fell down the stairs," she answered softly.

"What a clumsy person you must be, to have fallen down the stairs so many times." I sighed to myself. It would require a contortionist to fall down the stairs and sustain bruises only on the trunk and yet be bruise-free over the head and limbs.

At that point, I kept my peace. I sent her for X-rays and to the toilet to collect a urine sample. Her chest X-ray showed a fracture on her left fifth rib that had callus formation which meant that the fracture had not occurred in the last few days. Her urine sample tested positive for microscopic hematuria which meant that her kidneys were injured, likely her right one.

"You will have to be admitted for observation," I informed Ms Ti.

She gave a start. "I have to go home. I have a two-year-old son at home. The nanny is leaving soon. I just need some medication for my gastric pain," Ms Ti said.

I looked at her, scrutinizing her face. She refused to meet my eyes. It was a pretty face: fair skin, big brown limpid eyes, straight nose, and well-shaped lips.

"I have to go home!" she suddenly exploded, and abruptly stood up.

I held her left forearm, gently restraining her. All fight left her, and she sat down as abruptly as she had stood up. There was now a lost look on her face. I gave her time to compose herself.

"How did you get these bruises?" I asked softly as I held her left hand in mine.

"My husband," she whispered. I waited. "My husband hit me."

I tightened my left hand.

She had no tears only a lost look.

"He said the house was not clean," she whispered. "He said I dressed like a prostitute."

It went on for ten minutes - all the reasons for which she had been beaten.

This was 1994. Domestic violence was not considered a societal issue; it was considered a domestic one. There were no formal channels to refer these victims to, and the truth was healthcare professionals were not looking for it.

Ms Ti was the first case of spousal abuse I picked up who had not come to emergency department announced as a victim. There probably had been other victims like her whom I had not recognised. After Ms Ti, I was on the alert. Patients with a black eye giving a history of walking into a wall; patients with imprints of objects on their necks, limbs, or any parts of the body; patients with no obvious problems – alarm bells would ring in the background.

We also had to recognize that a perpetrator may accompany a victim to the hospital to seek medical assistance. I encountered one who, despite being politely asked to wait outside the consultation area, insisted on staying with the victim. He turned aggressive, took down my name, and threatened to report me to the minister for being rude. We had to call security, and finally the police, to manage him. He refused to stay away from the victim.

It was generally difficult for the police to step in twenty years ago because many of the victims did not want to make a police report. The prevailing sentiment then was that domestic violence was a domestic issue.

I remember when I started to look into this problem and did a press interview on it. I received a shelling. This shelling was not from my hospital management but from someone outside who said that he was representing the ministry. I sat in his office for one hour. It was pointed out to me that I had been indiscreet in discussing this issue with the press. At the end of the bewildering hour, I was informed that it would affect the country's Beri-Beri standing. I left his office even more bewildered, wondering what vitamin deficiency had to do with domestic violence.

In those days, the medical social workers in the hospital had resource constraints. Fortunately, we found non-governmental organizations (NGOs) who were willing to try ways outside the box to help different individual victims of domestic violence. Victims who spoke English were matched to NGOs who had English-speaking volunteers; Muslim women staying in certain postal districts were matched to the help group that covered that district. What happened when we could not find a match, because our network was not extensive? We would then call around to ask for help. We were somehow lucky and would be able to locate an NGO willing to bend or break their rules.

The concern of a few individuals and NGOs, and their dedication to helping victims of domestic violence, has seen a change in the attitudes of officials and society towards this domestic issue.

Unfortunately, we were not able to help Ms Ti. She left the emergency department after signing an "against medical advice" (AMA) form. It would have been a different ending if she had been attended to today.

Interlude I

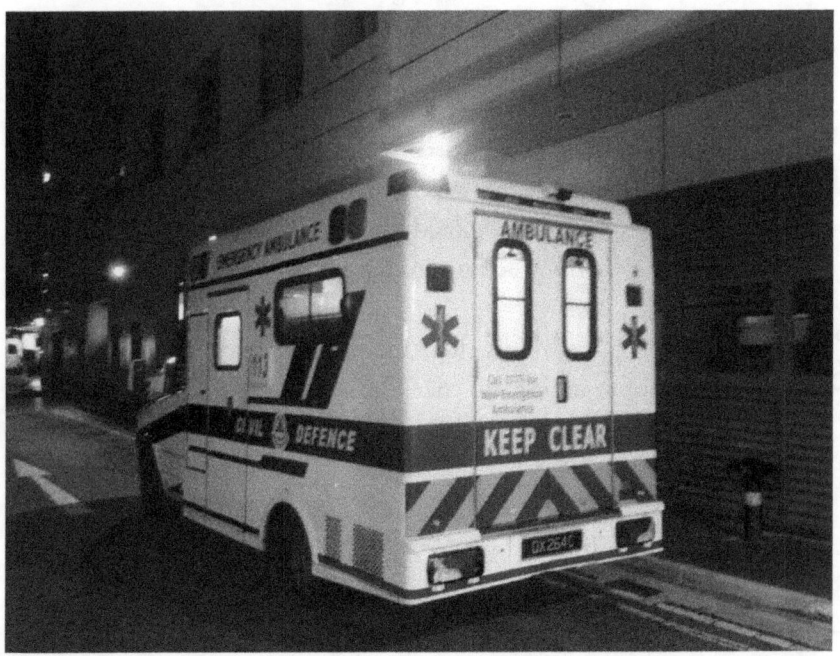

Resus – The Place, the Happenings, the People on a Particular Night Shift

It happened all in the space of ninety minutes. One moment, the resuscitation area was empty, and the next it started to fill up.

The first patient who was brought in, a thirty-year-old Malay gentleman, was parked in resus bay 4. He had called for an emergency ambulance because he had been feeling giddy. He also complained of having palpitations. The cardiac monitor showed ventricular tachycardia. His blood pressure was 70/50, which was below normal. He was in shock. We established two intravenous lines.

The patient was conscious. We explained that we were going to give him medications to sedate him, after which we would deliver an electric current to his heart to shock it back to a normal rhythm. He nodded.

The second patient, a ninety-five-year-old Chinese lady, was brought in as we were preparing to shock the first patient. The old lady was looking curiously at my fellow consultant, who was busy turning the knobs of the defibrillator. She was parked in resus bay 3, opposite the first patient. There was a screen between the two patients. This ninety-five-year old patient had been waiting for a hospital bed. She had been diagnosed with pneumonia, and treatment had been started earlier on. During their two-hour rounds, the nurses in the observation room had noticed that this patient's blood pressure had dropped and was below normal, 85/50. She was conscious and alert.

I left my fellow consultant and registrar to supervise the medical officer, who was managing this old lady. We ran fluids and made a presumptive diagnosis of septicaemic shock. At the same time, a twelve-lead electrocardiogram and a blood test were carried out to see if her low blood pressure was due to her heart. The electrocardiogram did not show any abnormality or acute changes. The blood test would be ready after twenty minutes.

My fellow consultant and registrar delivered the first electrical shock to the thirty-year-old Malay gentleman's heart. His heart returned to a normal rhythm, but a few minutes later the cardiac monitor showed that his heart was in ventricular tachycardia again. His blood pressure was back to 70/50. The team delivered two more electrical shocks to his heart. Each time his heart would return to a normal rhythm for a few minutes, but then it would revert to the malignant rhythm. We could not raise his blood pressure above 80 systolic despite the addition of an anti-arrhythmic drug.

The third patient was brought in as the third electrical shock was being delivered. Fortunately, this was a stable fifty-year-old Chinese gentleman who had pneumonia and heart failure, and he had been found to have a non-ST elevated myocardial infarct (NSTEMI) as well. He was brought in to the resuscitation area because he needed to be monitored continuously while waiting for a bed in the high-dependency ward.

Within five minutes of this fifty-year-old gentleman being parked in resus bay 5, the triage nurse pushed a ninety-year- old Chinese lady into resus bay 6, opposite him.

By now, the second patient, who was still very alert despite her low blood pressure, had received one litre of fluids, and her blood pressure had inched up to 95/55. I decided to start this patient on an inotrope, dopamine. This required a few minutes to prepare because it had to be given through an infusion pump.

The fourth patient, the ninety-year-old Chinese lady had been moving her bowels and had noticed that she had passed fresh blood together with her stools. She was feeling giddy and was pale as a sheet. Her blood pressure was 80/40 and her heart rate was 100 per minute.

The medical officer managing the fifty-year-old gentleman in resus bay 5 also took over the management of this ninety-year-old Chinese lady. We were running out of doctors. The entire senior team was in the resuscitation area together with half of the medical officer numbers.

As the fourth and fifth electrical shocks were being delivered to the Malay gentleman, others prepared to transfer him to the cardiac catheterisation laboratory for overdrive pacing.

The cardiology and medical on-call doctors were now in the resuscitation area when the fifth patient was pushed in. This was a twenty-three-year-old Indian lady who worked as a cleaner in a private hospital. While on duty, she had drunk a degreaser. She had burns on her neck and upper chest, a pattern consistent with liquid having spilled from her mouth.

It had been forty-five minutes since the arrival of the Malay gentleman.

The blood test of the ninety-five-year-old lady was finally ready, and her cardiac enzyme was raised. Her low blood pressure was due to a heart attack which was likely precipitated by her pneumonia. The medical on-call doctor agreed to move her to the medical high-dependency ward. This patient, who was very sharp for a ninety-five-year-old, was the first of the five patients to be moved out of the resuscitation area.

Because the cardiology on-call doctor was in attendance, my fellow consultant attending to the Malay gentleman took over the care of the twenty-three-year-old Indian lady directly (without a medical officer) from the paramedics, who had brought her in and parked her in resus bay 2. This patient's supervisor had brought the can of D9 degreaser, and a quick check on the Internet confirmed that it was a corrosive. We could not ascertain whether it been an accident or a deliberate act because the patient could not talk and was in distress; she was drooling and visibly in pain. We called for the on-call ENT doctor to clear the airway.

The Malay gentleman was still unstable with a blood pressure of 70/50 when we moved him out of ED to the cardiac catheterisation laboratory, one hour after his arrival in the resuscitation area.

At this point, the on-call ENT doctor was still uncontactable.

The general surgical team was reviewing the ninety-year-old lady who had bled per rectum. Her blood pressure had risen to 105/60 after 500 millilitres of fluids. A health attendant was on the way to the blood bank to get one unit of blood for her.

The on-call ENT doctor arrived fifteen minutes later as the fifty-year-old gentleman, who fortunately had an uneventful stay in resus bay 5, was moving out to the cardiology high-dependency ward. The on-call ENT doctor scoped and noted that the area around this twenty-three-year-old Indian lady's larynx was erythematous. Her airway had to be secured. This meant that this patient would have to be intubated as soon as possible, before her larynx became swollen. This had to be done in the operating theatre just in case the intubation attempt failed. If we failed to intubate the patient, we would have to proceed to a surgical airway, a tracheostomy.

There was another flurry of phone calls as the on-call anaesthetist had to be contacted to perform the intubation. We readied an operating theatre with equipment for a surgical airway on stand-by, and the general surgeon was ready to do an emergency esophagoscope, a gastroscope, and insert a feeding tube.

The general surgical team had escorted the ninety-year-old Chinese lady who had bleeding per rectum to the emergency operating theatre as the on-call ENT doctor was looking at the twenty-three-year-old Indian lady's throat.

We moved the Indian lady out as soon as we received a call from the operating theatre that they were ready to receive the patient.

Finally, ninety minutes after the Malay gentleman had been pushed into the resuscitation area, it was empty of patients again. The housekeeping team moved in swiftly to clean the five resus bays. We did not know when the next onslaught of patients would arrive.

Part 2

HAPPENINGS

CHAPTER 5

THE PARANORMAL

The Chinese believe that the seventh lunar month is when the gates of hell are opened and souls wander the earth. Offerings are made, and getai stages, or puppet shows, are provided to entertain them. Auctions are conducted, and paper money, paper gold, and paper clothes are burnt as offerings to these wandering souls.

The hospital is generally quieter during this month. The emergency department would see fewer patients, but every three to four years, there would be a freak rise in the number of patients. It is not possible to predict which year there would be such a rise or in which year there would be unusual happenings in the seventh lunar month.

I was told this happened in the old building during a seventh lunar month. At about 0330 hours, one of our senior nurses, Mr Tan, was in the pantry sleeping when he felt someone strangling him. He tried to pull the hands apart, but every time he pulled, they tightened like a python constricting around its prey. He started to panic. He knew he was asleep and had to wake up; he had to swim out of this suffocation. He was suddenly awake, in a cold sweat and gasping for air.

A minute later, Sally walked into the pantry. "Are you okay?" she asked as she noted his ashen face. When he did not answer, she poured a glass of hot water for him.

Over the next few days, Sally noticed the hypocount machine (it measures blood sugar) missing from the triage area. This happened a few times. She mentioned this to Mr Tan when she saw him.

"I will go and burn some offerings this evening," Mr Tan said.

The next morning, Sally was at the triage area, and the hypocount machine went missing again. She burst out in frustration. "Hey, we burnt the offerings yesterday. Why are you still playing around?" She looked around the triage area again. This time she found the hypocount machine on the table.

(Many years later, when Mr Tan recounted this event, he informed us that he learnt – he did not tell us how — it had been the ghost of a young child who had been up to mischief by hiding the missing hypocount machine)

Odd events also occurred in the next building and in a few places.

One of our Indian nurses saw Ox-Head and Horse-Face in the resuscitation area a few times. Ox-Head and Horse-Face are guardians of the underworld in Chinese mythology. They escort the newly dead to the underworld.

In another year on the seventh lunar month, we heard that a Caucasian nurse manager had seen a head floating around the intensive care unit during a night round.

When the ED team had to borrow a ward that had been mothballed for a few years, Sally was assigned to manage the area. After a few weeks, as Sally was walking around the new area, she saw a lady in white in one of the rooms. She asked her staff who the lady was. They stared at her. Sally thought her staff had not heard her question.

"The lady in that room," Sally said pointing in the direction of the room. "Who is she?"

"Sally, that room is locked. We have not opened it for use."

Sally was shaken. She went to the room and there was no one there. Over the next few days, she would sometimes see the lady in white in the room. No one else saw her.

A friend gave her a Christian prayer and Sally searched the Internet for more prayers. After saying them, she stopped seeing the lady in white.

Recently, a colleague had been avoiding our pantry during his night shifts. He had found himself waking up on the floor when he remembered lying down on the sofa to sleep. This had occurred a few times. No one else had the same experience.

There are areas in the ED where "sensitive people" will avoid lingering. Most of us know where these areas are, and we have learnt to work in them.

In previous years, patients would avoid the hospital unless they had a life-threatening condition. However, in recent years, this belief has diminished; more Chinese patients are having elective surgeries in this month.

Not all odd events occur on the seventh lunar month.

Many years ago, the health attendants from the ED were responsible for moving the bodies of patients who had passed away in the hospital wards to the mortuary. There would be two of them each time to carry out this duty.

On one of these occasions, Jafri and Hary had just moved the body of an elderly man to the mortuary. Jafri was the first to leave and did so without a backward look. Hary was not so quick; he heard the sound of a running tap and looked around. He noted that one of the taps was turned on, and he walked over to turn it off. He turned off the tap, but as he turned to walk away, he heard the sound of a running tap again. He stopped and saw that the tap he had turned off was on again. He stretched out his right hand to turn off the tap again, and as he did so, he felt a slap on his right hand accompanied by a resounding noise.

"Ouch!" he exclaimed as he withdrew his right hand. It was painful and he noticed a reddish mark on the back of his right hand. He ran out of the mortuary as fast as possible leaving the running tap behind. He was pale and in a cold sweat when he reached the ED. He was on medical leave for a week after this incident.

Something similar happened to one of our senior nurses, CM. The incident that Hary encountered occurred in the day, but this happened during a night shift, in the room we used as a minor operating theatre. CM was tidying up the room after a series of procedures had been completed. She had coiled tubing around an oxygen flow-meter on one of the walls and had noted that the gauge of the flow-meter was zero. Kate, another nurse on night duty, entered the room to inform CM that their supper was ready. Before they could leave the room, CM heard a hissing sound. She turned and noticed that the flow-meter which she had checked a minute ago had its gauge at ten instead of zero. She thought that she may not have switched off the oxygen, although she thought she had. Kate witnessed CM turning off the oxygen, and as they were about to open the door, they heard a hissing sound again. They looked at each other in shock, eyes wide. They exited the room without a backward glance.

Within that year, CM had an encounter with another piece of equipment. It was another night shift. After the handover from the afternoon to the night team was over, Mr Lee, an older colleague and a good friend, showed her a rather odd piece of paper. The paper was white except for a black border all around its edges, like a photo frame. Mr Lee said that he had heard the photocopier move although there was no one near it. He had gone to check and had found this piece of paper. CM laughed the incident off, assuming that one of the younger nurses was playing a prank on Mr Lee, who was known to be a sensitive soul.

CM and her team were busy in the first few hours of that shift. At about 4.00 a.m., there was a lull in activities. A group of nurses were hanging around the consultation bay opposite the treatment area, where CM was clearing equipment that had been washed. The same photocopier that Mr Lee had mentioned before was to her back. Another group of nurses were just around the corner, exchanging news about their colleagues. No one was near or paying any attention to the photocopier when they heard it move. All conversations ceased. There was an audible silence.

CM turned and asked the group of nurses at the consultation bay if any of them had been using the photocopier. All shook their heads, and none wanted to go near the photocopier. There was a visible piece of paper, and CM went to pick it up. When she turned that piece of paper over, it was a replica of the one Mr Lee had shown her earlier in her shift. The paper was white except for a black border all around its edges.

Both groups of nurses noted the pattern on the paper CM had picked up from the photocopier. A few were visibly shaken, a few had ashen faces, and some were nearly in tears. Both groups dispersed quietly and no one was in the mood to talk for the rest of that shift.

The next incident involved a trauma victim. When trauma victims are brought in to the resuscitation area and are unconscious, access is very important and is time sensitive. The team would cut patients' clothes at the seams and remove them. The patients would be examined for injuries from head-to-toe.

On a particular afternoon shift, CM and Mr Lee were together in the resuscitation area when an unconscious Chinese gentleman in his early forties was brought in by the emergency ambulance team. He was a motorcyclist who had been hit by a car. They cut his batik shirt along the seams; removed his belt, shoes, and socks; and pulled off his pants. He had bruises on his face, chest, and abdomen. He was intubated, and they started chest compressions because he had no breathing and was pulseless. After one hour of futile efforts, the team pronounced him dead. CM and Mr Lee prepared his body. They dressed him in his clothes; as for his batik shirt they put it back again and taped the sides where they had cut it at the seams.

A week later, when CM and Mr Lee were working together in another afternoon shift, they noticed a moth flying into the resuscitation area. The pattern on the wings of this moth was exactly that of the batik shirt they had put back on for the motorcyclist who had died a week ago.

There is a belief amongst the Chinese that the soul of a departed will return seven days later.

These happenings do not occur often; most times they are seasonal. Some places are more prone to such happenings, and a few people are more sensitive than others.

One odd thing we have noticed. We have not come across doctors who had such encounters. Our conclusion is that doctors have encountered such 'happenings' but are reticent about discussing them, or the paranormal merely bypasses them, or they are too dense to sense these 'happenings'.

We leave you to decide whether or not these 'happenings' we described occurred.

CHAPTER 6

BOGGLE THRESHOLDS

"Er, can we not order xiao long bao?" Jeff, the medical officer, requested.

Jeff, Cherry, and I were at Swee Choon Tim Restaurant after an afternoon shift, for supper.

"Huh?" I asked in a puzzled tone.

"I am on afternoon shift tomorrow," Jeff answered sheepishly.

Cherry, the registrar and I looked up from the menus we were perusing.

"Why?" Cherry asked before I could utter "huh?" again.

"It's a bao," Jeff answered matter-of-factly.

Cherry and I were taken aback. Most of us would not eat bao just before or during our shifts, but not the day before — and it was the traditional bao we avoided not xiao long bao. Unlike the traditional bao which has a white skin and can be stuffed with different types of fillings, xiao long baos are about five centimetres or less in diameter with pork as the traditional filling. Xiao long baos are eaten hot, and the way to eat one is to bite a small hole on the side of the dumpling; suck out the hot soup, taking care not to burn the tongue; and then put the whole dumpling in the mouth.

Why do we avoid eating a bao before or during a shift? The word bao sounds like bao ke liao in Hokkien which means "take all" or "to cover", "to wrap". Unless we were prepared to have a deluge of patients, very ill ones, or many deaths during our shifts, we would avoid eating a bao anywhere before or during a shift.

Renee Haynes, a writer and historian who died in 1994, introduced the concept of the boggle threshold – the point at which a phenomenon is considered highly unlikely to be real, or the point above which the mind boggles. This point is subjective and will vary from individual to individual.

With regards to the traditional bao, this is a boggle threshold most of us who work in hospitals and emergency departments have. However, there will always be outliers. Two of my friends, Honey (a nurse) and Joyce (a doctor), were too fond of baos to be concerned about this boggle threshold. One of them would eat bao openly at the hospital canteen; the other surreptitiously ate them at her office desk, though ours was an open office concept. Both could be seen enjoying their bao during their shifts.

Back to the supper at Swee Choon Tim. Cherry and I had our piping hot xiao long bao, and Jeff avoided them.

In the 26 July 2014 issue of the *New York Times*, in an article entitled "Where Reason Ends and Faith Begins", T.M. Luhrmann gave examples of belief continuum and boggleness. Jeff, Cherry, and I had just demonstrated an intergenerational difference regarding where we drew the line in our belief continuums; we had different boggle thresholds.

Colours are an area where boggle lines are also drawn differently. Red features often in most of ours. The wearing of red briefs was a belief that I thought had died a natural death, because I had not heard about it for decades. I was wrong. Today's young doctors have continued the tradition of wearing red briefs when on call or on shift, for better luck. However, for a few of us older ones, we found ourselves very busy when wearing red T-shirts and would avoid this colour.

A few decades ago, green was the other colour we would avoid during shifts. In those days we wore our street clothes in the ED, unlike today where we wear scrubs as our uniform. Back then, my friends and I were young and did not believe in these superstitions. I had this green dress of which I was fond. It was a pleasant shade of green, it was straight, and I still had a figure that could carry off such a dress with hints of curves in the right places. It was also comfortable and easy to move in, especially when I had to do chest compressions. It did not occur to me till a friend (a guy) said at the start of a shift, "Um, you will be busy today...."

"Why?" I asked.

"Is this the same dress you were wearing two weeks ago, when we had the multiple pile-ups?" Kane asked.

It was. On that shift we had to manage five major trauma cases from two different road traffic accidents.

I was busy that day but not as busy as the day, two weeks before, so I was not convinced. I wore the same dress one more time the following week, and it rained patients. While having dinner with Kane after this particular shift, he remarked, "Are you convinced?" as he pointed to my green dress. I laughed. I put the dress into storage during that posting. A new boggle line was drawn.

* * *

It was not their time yet.

A common belief amongst most of us is that people survive because it is not their time yet. Some attribute it to guardian angels, the one above, good luck, good karma, or being a good person. A few would put these beliefs down as sheer superstition and attribute it to coincidences of the stars being aligned. Honey shared the following two encounters she had with patients when it was not their time yet.

"I had just brought in the cards from triage to place on the nursing counter," she said. "I don't know why, but I decided to check out the area. All the consultation bays were occupied, but the curtains were not drawn, so I could see the doctors attending to patients and accompanying persons. The curtain in the area behind the nursing counter was drawn. I called out to ask if I could enter and assist. Dr Theeka was attending to the patient, and he allowed me to enter." Dr Theeka was an experienced and independent practitioner who required minimum assistance from us. "The patient he was attending to was in heart failure. His plan was to give 100 mg of frusemide. Frusemide comes in brown vials of 20 mg each. He had opened five vials and placed them on a dish. He had not drawn them into a syringe yet."

At this point, I was getting worried, although this happened many years ago.

Honey continued. "I really don't know why I decided to check out that area that day, but I'm glad I did. The five brown vials he opened were not frusemide."

I looked at Honey expectantly. In my mind, I remembered another incident that involved brown vials; that time it was frusemide and adrenaline. Someone had given one vial of adrenaline to a patient with heart failure instead of frusemide. Fortunately, there were no ill effects. Adrenaline accelerates the heart rate and can cause ill-effects in a failing heart.

I asked anxiously "was it adrenaline?"

Honey shook her head. "It was promethazine. Luckily I went in to assist. I pointed out to Dr Theeka the vials, and his face turned white. You know he is an Indian with dark skin, but at that moment when he realized that the vials were promethazine and not frusemide, his face looked white. Can you imagine if the patient had been given five vials – 250 mg of promethazine? He could have died from it."

Promethazine is a drug given for drug allergy and it can result in a drop in blood pressure when an overdose is given — 250 mg is an overdose. It is meant to be given via the intramuscular and not the intravenous route. Frusemide is generally given through the intravenous route which was what Dr Theeka had planned to do.

Honey's second encounter involved Olive, a medical officer who always reminded her of Olive in the cartoon, *Popeye*.

"I was not impressed with Olive initially. She moved very slowly, and it seemed that she thought very slowly as well. There was this day when I walked along the ambulance driveway and passed by the resuscitation bay. The door to the driveway was open, and I could see Olive and Nurse CM. They were preparing to move a patient out. CM had placed the cardiac monitor at the foot of the patient trolley with the monitor facing outwards. It was not facing Olive who was at the head of the trolley. I could see it as I passed the door. Olive was adjusting the oxygen mask on the patient. I glanced at the cardiac monitor as I walked past the door, and I did a double take and quickly walked back. The cardiac monitor had a bizarre pattern — the patient was in ventricular tachycardia. I shouted, and we pushed the patient back into the resuscitation area. The foot of the trolley had just protruded out of the door. The patient had no pulse. We defibrillated the patient, and Olive coolly intubated and resuscitated the patient. My opinion of her turned 180 degrees. She did very well and was not slow. In fact, she was a pro. We became good friends after that. CM and I found out later that she was an anaesthetic trainee. I learnt never to judge a book by its cover from that incident."

To this day, Honey does not know why she walked through those areas at those times.

*   *   *

In this next incident, Dr Howe did not know why he did what he did. It was fortuitous that he did.

Mrs Lee heard the shutter to their shop coming down. It was 8.00 p.m., and they were closing for the day. She then heard a *thud* and the shutter coming to a halt. She ran out from the back of the shop and saw her husband rubbing the top of his head.

"Are you all right?" she asked anxiously.

Mr Lee nodded his head.

"I think so. I was bit careless," he answered.

Mrs Lee looked closely at her husband's head. Fortunately, there was only slight redness over the top of his head. They went home and applied an ice pack to Mr Lee's head.

Over the next three weeks, there was no change to Mr Lee's daily routine. He felt well except for this rushing sound in both his ears. This sound got louder and louder, and it was rather irritating.

A few days before he visited the emergency department, Mr Lee developed a headache on both temples, which disappeared only after taking one gram of paracetamol. However, the headache would recur after six hours. Mrs Lee was worried because her husband was scheduled for a business trip to China. A lot was riding on this trip such as whether their business had to be scaled up or down.

Dr Howe noted that Mr Lee was a sixty-year-old Chinese gentleman who walked confidently in to his consultation room, accompanied by a Chinese lady in her fifties. The couple related to him the events of the last few weeks and the reason why they were at the emergency department. Dr Howe examined Mr Lee. He examined his cranial nerves (looking for any deficit), his peripheral nervous system, and his fundi and tympanic membranes. He did not find any abnormality.

Dr Howe was now in a dilemma. There was no reason for him to order any investigation. The headache was likely due to stress, which the patient and his wife had alluded to, but Mrs. Lee was worried. On his own, Mr Lee would not have visited the emergency department.

Dr Howe decided to probe further.

"He doesn't get headaches. This headache he has is taking a long time to go off," Mrs Lee said.

"The only accident Mr Lee had recently is the shutter hitting his head? Is that right?" Dr Howe asked Mrs Lee.

The couple nodded in unison. Dr Howe studied their faces closely. Mr Lee was frowning and Mrs Lee looked worried.

Dr Howe sighed and stopped tapping his right fingers against his right thigh (a habit he did whenever he had to resolve a dilemma). "Mr Lee can be admitted and we can do a CT scan of his head," Dr Howe started to say and he had to hold up his right hand to stop Mr Lee, who was about to protest. "Or we can do the CT scan of your head, but you will have to pay $321 for it."

Mr and Mrs Lee looked relieved. "As long as I don't have to stay in the hospital," Mr Lee said.

Two hours later, Dr Howe received a call from the duty radiologist. "Is Mr Lee YS, IC number, S1234567F, your patient?" the duty radiologist asked.

"Yes," Dr Howe confirmed. He was on the alert — the CT scan must be abnormal for the duty radiologist to call.

"Are you in front of a computer?" the radiologist asked.

"Yes," Dr Howe answered as he accessed his patient's CT scan results.

"Mr Lee has a symmetrical bilateral acute subdural haemorrhage with mass effect," the radiologist reported.

Dr Howe shook his head in disbelief as he scrolled down the CT scan slices. There was a helmet of blood around Mr Lee's brain. A patient with such a lesion should not have been able to walk into Dr Howe's consultation room so confidently.Dr Howe thanked his lucky stars that he had not discharged the patient with advice. No one would have faulted him because the accident had sounded like a minor one; the patient had seemed well in the last three weeks except for the headache, which could have been attributed to stress, as alluded to in the history taken.

"Mr Lee, I am afraid you will have to stay in the hospital" Dr Howe said as Mr Lee was wheeled in to the resuscitation area on a trolley.

"I suspected something was wrong when the X-ray staff insisted I must be sent back on this trolley. Now that I am in this room…." Mr Lee said as he nodded his assent.

Mr Lee was seen by the neurosurgeon on call and was admitted to the high-dependency unit.

* * *

The ED and other medical teams have a collective belief continuum and similar boggle lines, although individual ones may differ a few degrees in latitude.

CHAPTER 7

ABOUT DEATH

We see death on most of our shifts, and as we grow in our jobs, we build armour around our emotions. At times we can come across as cold and matter-of-fact to relatives who have just lost their loved ones. I came to understand that this is a necessary stage in the maturity of a clinician. It is difficult to be empathetic or sympathetic, and then in the next minute turn and manage a resuscitation, cranky patient, or one who has a mere cough. This is the emergency department, where patients are varied in their acuity and age, from a well-walker to a desperately-ill patient, from a new-born to a centenarian.

Most of us learn how to manage our emotions as we mature. We learn how to remain functional in the face of grieving relatives and after facing grieving relatives, but the deaths of certain patients will still hit us very hard.

I remember a beautiful three-month-old baby girl who was well-formed, was a little chubby, and had cheeks that we would be tempted to pinch. An ambulance crew brought her into our resuscitation area. It was about 2.30 p.m. on a Wednesday. The baby girl had no spontaneous breath and no detectable pulse. The paramedic was performing chest compressions, and his assistant held an oxygen mask over the child's nose and mouth. We took over from them and passed an endotracheal tube through the baby's windpipe to deliver oxygen to her lungs. A nurse took over the chest compressions from the paramedic. One of the senior doctors tried to obtain a lifeline (inserting a cannula through a vein) but failed. Another senior doctor tried on the other upper arm but failed. Finally we were able to deliver emergency drugs when one of the senior doctors placed an intraosseous needle in to her right tibia (a needle into the bone marrow of her right lower limb).

We worked on the baby girl for three hours, but we failed to bring her back. There was a pall in the resuscitation area. When the parents and grandmother were informed that the baby girl had passed away, they were very quiet; there was no crying or screaming. We put the body of the baby girl in one corner of the resuscitation area and screened it to give the parents and grandmother the space and time to say their farewell. A few of the nurses cried quietly; one of them had just returned after her maternity leave. One of the senior doctors involved in the baby's resuscitation efforts had red eyes.

The hospital's care and counseling team was activated to support the parents and the grandmother. We were concerned because there had been no tears. It took time for the team involved in this resuscitation to recover their equanimity. Fortunately, they were the morning shift team who were due to go off duty.

I was not affected very much by this baby girl's death except for a fleeting sense of sadness. Each of us is affected differently. Generally we are shaken when the patients who die are of similar ages as our own (facing our own mortality), or those who have young children, or those who remind us of people we miss.

When we lose patients who were friends or colleagues, the grieving is in the open, and it is easier to comfort and support. It is when we do not know what will trigger the grieving that we have difficulties recognizing those who require support.

No one knows that whenever I see petite Chinese ladies in their seventies who lie quietly on trolleys in the resuscitation area, I have a lump in my throat. If they arrived conscious and then leave the emergency department for the mortuary, I struggle emotionally in the next hour, trying not to tear in front of the next patient or the rest of the team.

Sometimes, these emotions are held in abeyance when other ill patients have to be cared for, but later when frenzy activities have been dealt with, these emotions will surface. I have found that it is much better to accept that these emotions exist and acknowledge them. This acknowledgement returns my equanimity, and I can move on.

* * *

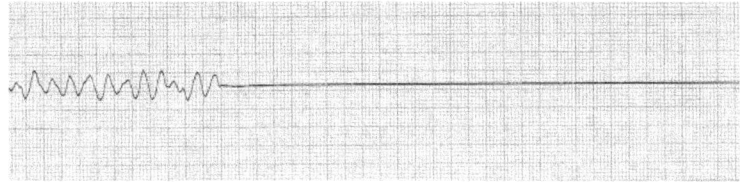

Death is not a common topic in dinner conversations, but it should not be a surprise to anyone that it is more common among healthcare professionals who encounter it in their work.

A friend recounted this incident during one of our catch-up dinners.

"The son was too calm for my liking when I told him that his father had passed away," Dan said. "It was not the usual response so I asked him, 'Are you ok? You are very calm'."

The son then shared this story.

The patient, Mr Tan, somehow knew that he was going to die soon. The day before, he had asked his son to spend time with him and keep him company. He had reminisced with him about bygone days, discussed his wishes, and shared his hopes and dreams for his family. Both the older and younger Mr Tans did not sleep that day. They talked through the night, and when dawn arrived, the older Mr Tan asked to go to his favourite bak kut teh stall for breakfast.

When Dan intubated the patient, he had to remove a piece of fat from Mr Tan's laryngeal opening. Mr Tan had choked and died.

Our table of doctors agreed that it was not a bad way to go – dying while eating a favourite food.

It is sometimes difficult for family members to let their dear ones go. Someone had mentioned before that patients who have doctors in their family are more likely to have an easier death. What does this mean?

Most times, doctors recognize what measures are heroic and accept their futility. However, we understand that it is difficult for people to accept that their loved one is ill and at the same time decide that only comfort care should be given. Every one of us requires time to reconcile ourselves to a possible loss. The expectation is for the loved one to be cured and to go home alive and well.

Rarely, I have encountered family members whose loved one had been ill for some time, snatched from the jaws of death before, and primed by doctors about the poor prognosis of the patient. These families would be prepared to let the patient go. When they were informed that their loved one has passed away, they would nod and say thank you with moist eyes. They would then politely inquire about what to do next, and most of them would have things ready.

The dinner conversation moved to what we would want done for us if we were ill. We all agreed that if we were dying, we would refuse to have a nasogastric tube (tube put through a nostril down the oesophagus to the stomach) inserted to feed us, to keep us half-alive. One declared that if she was to be fed and aspirate the food into the lungs, then she would accept that she may die from aspiration pneumonia. No heroic measures, please. No intubation, no chest compressions.

We then spoke about the age when we first encountered death in our lives. We agreed that the age when we first encountered death as a professional and in our personal lives did shape how we managed our emotions and our interactions with grieving families. One of our conclusions was that there was not much difference to our responses whether we belonged to the baby boomer generation, generation X, or generation Y.

The first time I attended a funeral was for a friend and schoolmate. She was ill for six months and then she was gone. Two weeks later, the wife of my supervisor became comatose and passed away a few days later. Two weeks after his wife passed away, *his* father passed away as well. I do not remember how my friends and I survived that month, but it did pass and nothing untoward happened to our patients. I lost touch with this supervisor, but decades later I was informed by a friend that he had died in his sleep.

The dinner conversation moved to the best way to die. Our table of doctors voted dying while sleeping.

I remember being taught in school to pray to St Joseph if I wanted a "good death". The definition of a good death then was to die in one's sleep. In the Catholic tradition, St Joseph is considered the special patron of the dying. Three reasons have been given: he is the father of Jesus, he is a terror to demons, and he died happily in the arms of Jesus and Mary, the mother of Jesus.

"It can be hard on the family when a patient dies in his sleep, but sometimes it is worse when a patient who was well suddenly drops dead," KP mused.

We looked at him, waiting for him to continue.

"They brought in this sixty-year-old jogger last week. No identification on him," KP said breaking the somber silence. "A fellow jogger saw him clutching his chest and then falling. That passer-by started chest compressions, but by the time the ambulance brought him in, he was flat."

There was silence around our table.

"An hour later, two ladies came inquiring about a man. They brought a photo of him and we referred them to the police." KP sighed. "Unfortunately, we weren't certain it was the patient. Later, when the identification was confirmed, one of the nurses went over to the mortuary to offer them assistance. That is why, all apologies should be made before going to sleep, and say I love you to those we treasure," KP concluded.

Eillyne Seow

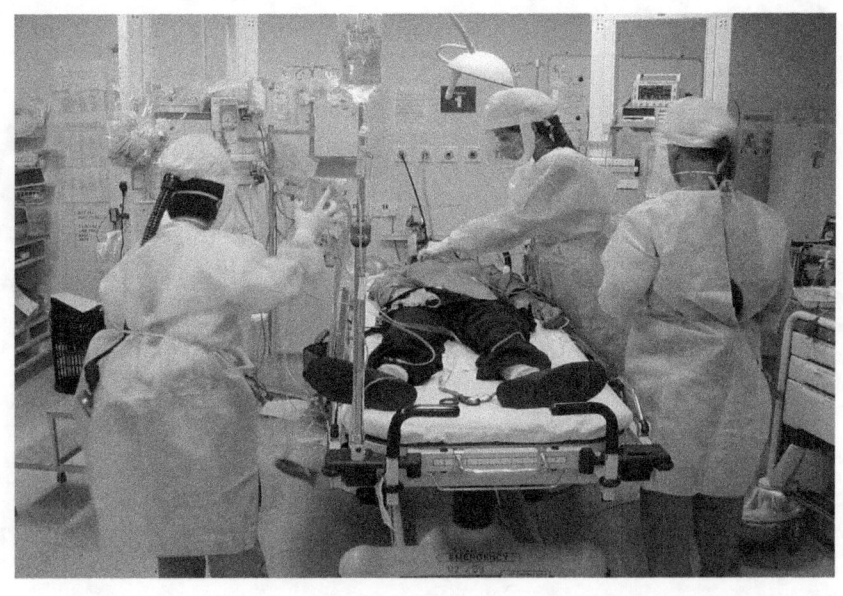

SARS – AN INFECTIOUS DISEASE OUTBREAK THAT WAS NOT EXPECTED

It began for me with a phone call. I was post-night (I had been on duty from 2230 hours the night before to 0800 hours that morning). It had been a memorable shift because my team had to deliver a baby in the early hours of the morning. The mother was a fifteen-year-old girl who did not know she was pregnant and neither did her mother. Deliveries in our ED are not common events.

Back to the phone call. It was from one of my senior doctors, SY, who informed me that the doctor in charge of infection control, Dr B, had been to ED to warn us that there were clusters of hospital staff who were ill with similar symptoms and signs. SY had discussed the situation with her, and it was agreed that the affected staff would be instructed to report to the ED. SY had called to inform me of their plan and to obtain my concurrence, because I was the head of the ED.

Fortunately for us, there was a separate building just next to our ED. This building was built to manage casualties from hazardous materials incidents should they require decontamination. It was equipped with showerheads and an appropriate drainage system. We called it the decon

51

building. According to SY's and Dr. B's plan, the affected staff would be isolated and attended to in the decon building. It was also decided by SY and the senior members of the team on that shift that only senior and experienced doctors and nurses, would attend to these patients. I felt proud of SY and the senior staff on that shift who consciously made the decision to send only senior and experienced staff to the decon building. SY explained that it was safer this way.

The doctors and nurses going into the decon building wore an N95 mask, full surgical gown, and gloves; at that time it was considered full personal protection equipment. It would later include a cap, goggles and shoe covers as the SARS outbreak progressed.

This was also the evening of our hospital's dinner and dance. I was told that the senior management and infectious disease specialists were rather preoccupied during the event.

The next day was the beginning of the daily 0800 meetings that would last till Singapore was declared SARS free in May 2003. Those meetings were interesting times for me not only because I was watching a national event unfold real time, but also because of the responses of the hospital leadership and my colleagues. These meetings were chaired by the chief executive officer and attended by the heads from critical areas like hospital operations, logistics, human resource, emergency planning, epidemiology, nursing, laboratory, and clinical departments. The mood was sombre most times and became worse when the number of deaths started climbing.

Looking back at those daily meetings, I had the impression that the CEO was firing at me in the first six weeks of the SARS crisis. It was a good strategy because it kept everyone alert, even those sitting on the sidelines. I went into each meeting expecting to be fired at and was mentally prepared for that. Once when it was flagged that a medical certificate of more than the officially stipulated number of days had been issued from the ED, the CEO turned to me furiously and I was the focus of attention for the next three minutes. It was a long three minutes for me.

I reported later that the medical certificate had been issued by a doctor from another department, not from the ED.

After these 0800 meetings, I would return to the ED to update the team and implement changes. Fortunately, WhatsApp, Facebook, and Tweeter were not the norm then, but even so, changes to our workflow occurred two to three times a day. One of the more difficult changes we had to make was ensuring that our team complied with infection control measures 100 per cent of the time. I was very frightened that I would lose a member of my team to SARS. The only defence against SARS which we knew of was infection control.

We were attending to patients in the sweltering heat outside the air-conditioned building. By the end of March 2003, we were the national screening centre, and most of our work had moved to the decon building and the area outside our main building. This area was hot. Have you ever parked your car out in the open? When you get into the car an hour later, it feels like a sauna. That is how it felt working under the tents we put up outside the ED. Then imagine working under the tent and wearing a cap, goggles, N95 mask, surgical gown, gloves and shoe covers, and attending to patients for eight hours. One of our staff said that breathing through an N95 mask was like breathing through a piece of cardboard which smelt of urine. Others had to bear with the pain on their nasal bridges where a metal piece was attached on the mask. This metal piece had to be moulded to fit the shape of the nasal bridge for the mask to fit well. The mask had to fit well to prevent leaks. Another problem for a few of the ladies were pimples along the outline of the mask.

The screening area expanded outwards, forwards, and sideways till it occupied the small car park in front of the ED and a third of the car park to the right of the ED. The decon was to the left of the ED. The configuration of our working area changed frequently as the numbers increased and the severity of the outbreak changed. I still remember those bright yellow chains that were used to demarcate the three groups of patients we screened (low risk, moderate risk or high risk); they gave a dash of colour to our working environment. Otherwise, we were working in a drab area which was funereal, which was to be expected because the materials used to build the screening area were the same as the ones used at wakes. The differences were that there were more water pipes and electrical lines than during a wake. I salute the contractors and their workers: although afraid of the SARS bug, they were still willing to come and do the business.

We were also creative in the way we cleaned the area. The at-risk areas were demarcated using plastic chains which were easy to clean. The consultation rooms (i.e., tent areas) were wiped down after every consult, conveniently done with mostly consumables which could be used, thrown, and replaced.

As the national SARS screening centre, we received patients from all healthcare institutions, primary care and the borders. Patients from the borders were the most interesting and put up the most resistance to being admitted and isolated.

There was a gentleman from Shanghai who arrived at Changi Airport and was noted to have a fever when he passed the temperature sensors. He was sent to us for screening, and because he was considered at risk of having SARS, we had to admit and isolate him. He raised a ruckus, shouting and throwing chairs around. The police had to be called in, but surprisingly when he saw the men in blue, he calmed down and meekly got himself admitted.

This disturbance led the two senior doctors on that afternoon shift, Jon and Wilson, missing their dinner. They left the ED about midnight, their stomachs rumbling, and so they headed for the coffee shop at the corner of Balestier Road and Mandalay for a bak kut teh supper. They sat at a table al fresco, along the kerbside. Jon sat facing the road, and Wilson had his back to it. A car came careening into them and the tables adjacent to theirs, scattering everything like tenpins on a bowling lane.

I found out about the incident the next morning when both came to report for work. Jon said that he saw the car drive towards them and was frozen in time, petrified. He did not have the presence of mind to warn Wilson, who took a direct hit. After the car had come to a stop, they noticed the driver running away from the scene. They thought the driver was not running in a straight line. Wilson and Jon checked on the rest of the casualties, ensuring that none required immediate medical attention. When the emergency ambulance teams arrived, they declined to be conveyed to the hospital (which would have been Singapore General Hospital although Tan Tock Seng Hospital was just round the corner; at that time Tan Tock Seng Hospital had stopped receiving non-SARS patients). The one thing that annoyed them was that they had to go home hungry.

We roped in the senior doctor, who was on his day off, to cover their morning shift. Both were rather stubborn about being examined for injuries but gave in to orders which is typical of doctors. Jon was fine, but Wilson had microscopic hematuria (suspicious of injury to the kidneys). I was rather shaken by the incident thinking how the team could have lost them in one accident.

During the SARS outbreak, I cried once in front of a few of my team members. This was early on during the outbreak, when there was very little information. Over coffee at Starbucks, as I was explaining to them why they had to be vigilant, I was overwhelmed with my fears and was not able to hide my tears. I had this fear that I would lose one of my team to SARS. When a friend from another hospital called during the SARS outbreak about one of his doctors, I could empathise.

In early April, I received a call from a friend who worked in another hospital. I was then attending to a patient in one of the tents under the ambulance entrance, and I was wearing personal protection equipment. I arranged for my patient to be attended to by another doctor. I had not been in touch with this friend for some time, and I did not expect it to be a "how are you?" call.

One of his doctors had a fever. At that point, all healthcare workers with fever were to report to us for screening. He was referring his doctor over to be screened. I suggested that the doctor pack a bag. My friend reacted very strongly to the suggestion. "Why are you jumping to conclusion that he has SARS? You haven't screened him yet!" I placated him and directed him to send his doctor, Dr Lim, to the car park in front of the ED. I wore personal protection equipment to meet Dr Lim at the car park, gave him an N95 mask, and walked with him to the decon building, where we screened our patients with high risk.

We were both sombre and, not in the mood for small talk. I attended to Dr Lim because I was the most senior on duty. Dr Lim was not convinced that he was at risk of having SARS. He commented that he had finished a two-kilometre run the evening before with no problem. I reserved my comments and did not want to upset him. I did not ask for a history of his illness or possible contact with SARS patients from him; I did not carry out a physical examination because he looked disturbed. Dr Lim was directed to the other end of the decon building, where we had cordoned off an area for shooting X-rays. This end was furthest away from the ED entrance, and portable lead shields were used to demarcate the area. A chest X-ray was arranged for Dr Lim. The chest X-ray showed a five-centimetre patch over his right lung field. I showed the chest X-ray to Dr Lim, and he was silent. We made arrangements to admit him to a general ward, because he was stable then. Later he would be transferred to an intensive care bed. Fortunately he survived.

There were other healthcare workers who were not so fortunate. This was the disease that made most of us feel vulnerable. Before SARS, we had this feeling that we were invulnerable to the bugs our patients carried. SARS humbled us. A friend in one of our in-hospital team lost a few of his team members, and he was devastated. That was the constant fear I had during the outbreak. Another fear we all shared was spreading the bug unknowingly to our family members. A few stayed away from home, and many changed before going home. Most of us could understand why members of the public shunned healthcare workers, but being treated like pariahs depressed many of us.

The irony was that it was initially difficult to convince the public that we had to restrict the number of visitors into the ED. Eventually when the deaths started to climb, the resistance crumbled. I read with amusement a newspaper article a few months ago about journalists risking their lives to cover the SARS story. I remembered how we were trying our best to keep them away and safe from the screening area, as well as protect the privacy of our patients. In the initial weeks of the outbreak, we would catch one or two a day using long-range cameras, but these intrusions stopped as the deaths mounted.

Another group that posed a challenge in the initial weeks of the outbreak was those who were on home quarantine. They were exposed to SARS patients but were well. We had not planned what we would do should they require admission to the hospital but did not have a fever. At that time, fever was the main reason patients would be isolated if there had been contact with a SARS patient. I was woken up at 2.00 a.m. on the day the first patient on home quarantine was brought into the ED. She was not well but had no fever. The working diagnosis was anaemia, and she was breathless. She required admission because she was symptomatic, and her hemoglobin was 7 g/dl (the normal range is 12 — 17 g/dl). The question was where to admit her to, general or isolation. In those days I was the gatekeeper, which was why I was called. I decided to admit her to an isolation room. The logic was she was on home quarantine and required isolation; therefore if she was to be hospitalized, she would require isolation as well, even if she did not fit the definition of a patient suspected of having SARS. It was fortunate, that we isolated this patient, because she turned out to have SARS.

The situation was fluid, and there were revelations and changes, sometimes more than once a day. We had decided early to confine decision-making in the ED to the eight senior doctors we had; this minimized confusion on the ground for us.

We had to think on our feet most of the time, and we had a philosophy of scrutinizing the definition of SARS, especially those given by international official bodies. After all, it was a new disease. One definition we took with a bottle of salt was that of contact history. In the early days, contact history (i.e., known to have been in contact with a SARS patient) was a criterion for suspecting that a patient had SARS. An alert general practitioner referred an extended family for screening because they all had fever, and the grandmother of the family had just passed away. An autopsy had been done, and the diagnosis of SARS had not yet been made. Unbeknownst to us, SARS had spread into the community. It was a Sunday afternoon, and I was clearing paperwork (we still had that during an infectious disease outbreak) when the senior doctor on duty walked into my office and informed me about this family. We looked at each other, nodded and said "SARS" in unison. It would have been a comical scene to any observer looking at us, but this cluster of patients was sad news.

Our epidemiology colleagues worked overtime to track down all those who had been in contact with the members of this family. They had been doing detective work since the beginning of the outbreak and had put in back-breaking efforts to track down every possible contact. There was one instance when they were recounting how they had to track down the "mahjong kakis (fellow players)" of a suspect patient, and the patient could only identify his fellow kakis by their first names.

The ED team worked very closely with them and the infectious disease team. Most of us have remained friends since then. A common saying is that the true colours of a person are revealed under stress. I would extend that further to true friends as well as enemies.

By the end of April, our colleagues in other emergency departments were having a worse time than us. They had to manage higher patient loads because we were closed to non-SARS patients, and yet they had to ensure that their patients were safe from other patients. Some patients who did not suspect that they had SARS attended these other emergency departments; and there were those who were in denial, afraid of attending the national screening centre. It was only human to be afraid.

Initially the staff of the hospital were shunned, but after this problem was highlighted by Dr Balaji, then the minister of state for health, there was an outpouring of support from members of the public. During the outbreak, many showed support in kind – LE Café gave the staff cookies, an anonymous donor drove by and gave several boxes of fruits, a bakery called Breadtalk donated unsold items to the night shift staff for supper, and the senior management was known to buy the staff food. During a lull on one particularly busy night shift, the acting chairman of the medical board came by the ED at 4.00a.m. to give food that he had bought for the staff on duty. He passed it to a nurse, who refused to accept the food and told him that she had to obtain the permission of the head of department before she could accept. The chairman was apparently so exasperated that he left it with her and just walked off. When she recounted the story, the whole team went into whoops of laughter. It was no wonder that the chief executive officer had described me as an ogre.

By mid-May we were eagerly anticipating that Singapore would be declared SARS-free and that life could return to normal. Our hopes were dashed a few times, but by the end of May we were officially out of the woods.

In the months to come, the staff would be interviewed about their experiences during those months, and these interviews were penned in two books.

I was happy to know that the team had faith and trusted their seniors, and that food had been a comfort in those hot days.

However, SARS did affect the mental health of the healthcare professionals. In the months after Singapore had been declared SARS-free, one doctor found herself crying over the potato salad she was making, and fortunately she had the insight to seek help. Another was brought in drunk by the police to the ED;

the staff member was physically and verbally abusive towards the senior doctor who was attending to him. It was a difficult encounter for both because the attending ED senior doctor, who had been through the SARS outbreak, could empathise with the emotions the patient was having, but he could not condone his behaviour. The patient returned a few weeks later - sober - to apologise.

Interlude II

Snapshots

It was a typical Thursday afternoon in the emergency department – slow paced.

My patient was a forty-year-old Chinese gentleman who sat quietly on the chair next to my consultation table. In my younger days, I was rather matter-of-fact with little niceties in my interaction with patients.

"What is the problem?" I asked Mr See.

There was a pause. I looked at him expectantly, my pen poised and ready to write on the case record. Mr See frowned and became very still. (This was the pre-Internet and the pre-Viagra era.)

"I cannot do it. My wife wants to divorce me," he said in an even tone.

Fortunately, I understood him the first time, and he did not have to repeat himself.

I nodded my understanding and recorded his complaint, which gave him time to compose himself. He was referred to a specialist managing such problems.

* * *

While having a drink with an old friend, as expected at our age, we started to reminisce about our past postings. He was an internist who, in his early years as a medical officer, had done a six-month posing in an emergency department. He shared the following.

I enjoyed my posting in the A&E [accident and emergency – the old name] very much. I was with David, and till today we are still in touch.

A&E is one place you get to see a different aspect of medicine. I remember I was on shift with David one night when a couple was brought in after being involved in a traffic accident. I attended to the girl. She was all right, but she was worried about her boyfriend and kept asking about his well-being. She was in her late teens and was petite with a very pretty face.

After completing her care, I went to check on her boyfriend. I stepped into the consultation room where her boyfriend was being attended to. It was then I discovered that the boyfriend was a girl, and my patient did not know. I did not meet my patient again, and I suppose she would have discovered this when both she and her boyfriend were admitted to the women's ward.

* * *

In the emergency department, we constantly guard against biases. We sometimes succeed, but sometimes we fail. It is easy to stereotype patients but hard to keep an open-mind.

One late afternoon, the emergency ambulance service brought in an eighteen-year-old boy. There was a strong smell of alcohol as they whizzed him past our central counter.

"He started rather early in the day, didn't he?" I commented.

An hour later, I asked the triage nurse, "What did the boy drink, and what time did he start?"

The triage nurse laughed.

"The boy splashed himself with cheap cologne. He is very sober," she informed me.

I laughed as well.

* * *

"Why? Why? Why?" I asked.

We had been exchanging stories about the weird things patients do, and Francis had just described the most graphic in his collection.

"How would I know why he stuck an iron rod through his anus? You are just like the MO [medical officer] who saw him," Francis exclaimed in exasperation. "My nursing officer was so annoyed with the MO who was attending to the patient that she came over to my consultation room and told me to quickly settle the management of the patient."

"Huh?" I asked blankly.

"Apparently the MO, like you, had been asking the patient, why for an hour," Francis said laughing away.

I was stunned silent for a moment, and then I burst out laughing as well.

* * *

Two patients, a motorcyclist, and his pillion rider were brought in to the emergency department after their motorcycle had skidded on a wet surface. The motorcyclist was a gentleman in his mid-twenties. He was examined and was fine.

The pillion rider was a pretty girl in her early twenties. She was wearing a leather jacket, which she refused to remove. We required her to remove the jacket for us to examine her. She was shy and kept still for about twenty minutes despite us coaxing her. The young girl finally relented. When she unzipped her jacket, we realized why she had been reluctant to remove it.

She was topless.

* * *

Part 3

PEOPLE IN THE ED

CHAPTER 9

A CONVERSATION: VVIPS, DOCTORS, AND PATIENTS

The emergency department is a great place for people watching.

All strata of society, including VVIPs will visit the ED sooner or later as a patient, as a relative or friend of a patient, or as a good Samaritan. Let me share with you my observations on a few of them.

VVIPS behave in two distinctive ways. We had a few who, as patients or accompanying patients, waited quietly together with everyone else. There was one, who despite running a temperature, spent the night chit-chatting with his fellow patients. The system alerts the staff to the presence of VVIPs, but occasionally the staff fails to notice the cue.

On the other hand, we have encountered the other group, whose arrival is announced even before they are in the ED. There was one who, even before registration, declared himself to be a VVIP, but unfortunately he drew blank looks from the staff. He was visibly upset with the response.

There were times when the name of a VVIP was mentioned, and the staff would ask, "Why is he a VVIP?"

Another group is the name droppers. To be fair to this group, they generally do not name drop unless they cannot get their way.

We had a rule that allowed only one person to keep each patient company. One of the patient's sons wanted an exception to be made for him. When the doctor did not accede to his request, he threatened to report the doctor to the hospital's CEO. He also boasted that he was on close terms with a certain minister and frequently had coffee with that minister. Unfortunately for him, the doctor he was threatening was very well-connected.

Apprehension is the most frequent emotion encountered in our patients and those accompanying them. No one wants to be in an ED except for our regulars – more about them another time. This emotion can give rise to unpleasant and difficult encounters, especially when the one accompanying the patient is a doctor.

Doctors as Relatives – The Worst and the Best

As a relative accompanying a patient, doctors are the worst and the best. They are the best when they know what the problem is, because they understand the rationale behind the possible solutions. They know that there are differing risks and benefits involved among the different solutions available.

A doctor's father was brought in with low blood pressure and was found to have an acute myocardial infarct (AMI). This was not a fresh myocardial infarct from the history, it had happened three days ago. The patient had seen a general practitioner for nausea the day before but had not informed the general practitioner about his backache and breathlessness. The general practitioner did not make the diagnosis. The patient, who had diabetes mellitus and hypertension, had waited for his son to return from abroad. The son took a look at his father and sent him to the emergency department.

At the ED, a decision had to be made whether the patient should undergo percutaneous catheterization (PCI) immediately or wait, because it was a Day 3 AMI. The son made the decision to go ahead with the PCI. It took only a few minutes for the son to arrive at a decision. His colleagues could put across the risks of the procedures in a clinical manner baldly, without having to sugar-coat the numbers. The son understood that the statistics being presented to him were pooled statistics that is, if it were 1 per cent mortality rate, it would mean that out of a hundred patients, one would die. However, if that 1 per cent were to be his father, then the procedure would have a 100 per cent failure rate for his father.

This knowledge can also paralyze doctors when their loved ones are involved. They know the worst and can imagine it happening. They sometimes cannot stop themselves from interfering in the medical care of their loved ones.

The father-in-law of a doctor from another hospital was brought in by the emergency ambulance team. The son-in-law, who was an intensive care specialist, barged into the resuscitation area and wanted to take charge of the care of the patient. When asked politely by the nurses to wait outside, not only did he refuse, but he started to shout orders at them. He left the resuscitation area after the senior ED doctor shouted, "Get out! This is *my* resuscitation room. Wait outside!"

We understand and empathise with the emotions relatives go through. After all, a few of us have been there before.

One doctor was in the ambulance when her mother stopped breathing and had no pulse. We did not have the heart to ask her to wait outside as we intubated the patient and did chest compressions. She looked stunned and lost.

Ten years later, when we encountered this doctor again, she was extremely obnoxious. This was her second time to this ED. This time her niece, who had fractured her ankle, had been brought in by the emergency ambulance team. The most senior doctor on duty had attended to the girl because she was twelve years old. The woman barged in to the ED and demanded that the patient be transferred to the private hospital where she worked. She wanted a copy of the X-rays. Hard copies had to be made because the hospital had gone digital, and this took time. She was not happy with the delay because she wanted things done immediately. The staff had other patients and priorities. The staff went by the book because catering to her request would be unfair to others. It did not help that she was rather hoity-toity with the staff in general. She was in a huff, pacing up and down the consultation area, and was entering consultation rooms and using the phones there without a "by your leave". I overheard her mutter under her breath that she had patients waiting for her in the clinic. She became more circumspect when she saw me. She recognized me, but I had a bit of difficulty placing her. We managed to find a lull in our activities and expedited her niece's transfer before a few of our team members could give her a piece of their minds.

"Hi, do you recognize me?" a doctor said to me during a tea break at one of our annual scientific meeting. I must have looked puzzled because I did not recognize the speaker. After she mentioned that her father had COPD (chronic obstructive pulmonary disease) and had been under the care of Professor SW, one of my former consultants and teachers, I then remembered who she was. This doctor, Dr Wong worked in a community hospital. Whenever her father had an exacerbation of his COPD, he would be brought in to the ED for stabilization and treatment before being admitted. Dr Wong would have started treatment for her father at home, and only when he did not respond to the treatment would she send him to the ED. She would be highly strung by then and would be speaking in a high-pitched and shrill voice. Her behaviour was not the worst that we have encountered, but as a general rule, a more senior nurse, doctor, or I would attend to her father. Most times she would calm down when she saw a familiar face. It had been five years since I had seen her. She informed me that her father had passed away five years ago.

"Thank you for helping me when I brought my father to the A&E. I was a difficult relative to handle. I must have been a pain," she said.

I was rather stunned and could only make some polite noises. This was not the first time I had received thanks for helping doctors, but it was the first time a doctor had acknowledged that her behaviour had been less than polite - and had thanked me for managing it.

Doctors Can Be Misers

There were three of us having coffee at a Starbucks café near the hospital. It was late afternoon and we had just finished a busy morning shift.

"The GP finally agreed to pay for the CT head scan?" one of my coffee-drinking companions asked.

I nodded as I put down my cup of latte.

It had been a vexing incident. A radiologist had brought in to the ED his friend's wife after a head injury. His friend was a GP. One of the ED consultants had expedited the process for this patient. The CT scan for this patient was completed even before the admission process was over.

This was when the problem started. The GP refused to pay for his wife's CT scan head.

In the usual case, the CT head scan would have been ordered and completed only after the formal admission process had been completed. The CT scan would not be an out-of-pocket expense.

In this instance, the CT scan head had been read by the radiologist as normal and the GP wanted to bring his wife home. This was fine, but the GP refused to pay for the CT scan head. The excuse the GP gave was that he had not been informed he had to pay for it out-of-pocket by the counter clerk. We could not blame the counter clerk because she had not known that the CT scan had been expedited. The GP was told that there had been a miscommunication, but he adamantly refused to pay. This meant that the counter clerk would have to pay the cost.

FC, *one of our senior nurses,* made a comment. "Trust a doctor to do this kind of thing. Nurses would have paid." I had no answer for her.

There was collateral damage from this incident as well. The ED consultant who attended to this patient was at the receiving end of a deliberate 杀一警百 (Shā yī jǐng bǎi) i.e., "kill one, warn 100". This doctor received a blistering earful from his head of department, which was audible from one end of the office to the other. Their relationship deteriorated and was never the same again.

The GP paid for the CT scan eventually — after he was informed by the attending ED consultant that if he did not pay for it, then the ED consultant would have to do so. The GP apparently said, "That is not right".

"So it had to come to that," another of my coffee-drinking companions said with a shake of her head.

"Yeah, why did he have to be like that?" I said, also with a shake of my head.

Continuing the Conversation

What about doctors as patients?

Doctors know that fellow doctors are difficult patients, because most times doctors who are patients take risks with their own health and lives. They push the limits of what is safe care.

One of my hospital colleagues, Dr Chew, registered to be seen at the ED. A few hours later, the senior doctor managing him approached me for assistance.

Dr Chew was requesting to be discharged, but it was not safe to discharge him yet. His serum potassium level was 2.3 mmol/L. The normal range is 3.0 — 5.0 mmol/L, but there is a possibility of a laboratory error of plus or minus 0.5 mmol/L. In this case, even if there was an error of minus 0.5 mmol/L, Dr Chew's potassium level would be 2.8 mmol/L, still below norm.

Anyone with low serum potassium is at risk of having his heart beat in an irregular fashion, and the heart can even stop beating. Dr Chew was well aware that he was at risk of having a cardiac arrhythmia. Low serum potassium can give rise to another problem: muscle weakness sometimes paralysis. Dr Chew had muscle weakness although he was able to drag himself around.

Occasionally we manage young, seemingly healthy young men who present with weakness in both lower limbs, but before we label them as malingerers, one thing we do is check their serum potassium levels.

I went to see Dr Chew. He was adamant about going back before midnight. It was about 10.00 p.m., and our initial plan was for him to be admitted. He was ready to sign the AMA form. We compromised, and he agreed to stay for another two cycles of intravenous potassium before signing the AMA form. I knew that he stayed nearby and would walk to work. I asked him not to walk home when he was discharged; we would arrange for a taxi to bring him back. He nodded, but of course he walked home.

The Conversation Moves On

Another group of patients that can be difficult are university graduates who are in senior positions in their fields of work. They also tend to consult the Internet indiscriminately.

We had emptied our first cups of coffee after a morning shift and were exchanging stories about difficult patients.

One of them had been a patient who did not believe that she had diabetes despite her random blood sugar level being higher than normal. Another had refused to take a course of antibiotics despite an elevated white cell count.

Our hypothesis was that these patients command people in their daily lives, which made it difficult for them to *receive* commands from another person. We empathise: doctors in similar situations may behave in a similar fashion. These patients can be difficult, irritating, irrational, and exasperating, but they can provide a good laugh after the event.

We ordered a second round of coffee as I described the encounter I had with one quirky patient during a night shift.

The Quirky Patient

As expected, it was a busy night shift. My registrar was held up in the resuscitation area, one standby after another. My fellow consultant was busy clearing the queue, running in and out of the resuscitation area, and checking on the registrar.

I was busy sorting out the information my junior doctors were presenting to me on the patients to which they had just attended.

We were hoping to have an uneventful shift. Uneventful not in the sense of less patients — that would be wishful thinking – but in the sense that we could get on with the work, without having to deal with any melodrama. Well, it was really wishful thinking on that shift.

At 1.00 a.m., the ambulance deposited a fifty-two-year-old woman who told the triage nurse that she wanted an anti-rabies injection. There were patients who were more ill than her who had arrived earlier and had yet to be attended to. The team worked steadily through the queue.

At 1.30 a.m., this woman approached the central counter and inquired politely, in her pseudo-American accent, when her turn would be. She was told politely that there were patients who had arrived earlier and had to be attended to first. She accepted the answer and headed towards the toilet. She was accompanied by her sister.

At 2.00 a.m., she approached us again to inform us that she had a colonoscopy in the morning. There was not much we could do because there were still patients who had arrived earlier, and a few were groaning in pain.

The ambulances continued to deposit more patients at a rate of five per hour, and walk-in patients continued to walk in.

The fifty-two-year-old with her pseudo-American accent and, permed, puffed-up hair, headed for the toilet again. "There is no toilet paper," her sister informed me.

"All right, I will get help," I reassured them. A staff nurse located the cleaner, who took care of the matter.

At 2.30 a.m., the fifty-two-year-old woman, with her pseudo-American accent and permed, but now not-so puffed-up hair approached the counter again. I was busy typing notes and checking on the junior doctors' work when she exploded. "I was bitten by a dog. I can die from rabies! It's critical also!"

I looked at her with a deadpan face, "We will see you, but we have other patients to attend to first."

She stared at me, turned, and walked back to her trolley.

At 2.50 a.m., a gentleman approached us to attend to his wife, who was in severe pain. The first available doctor was assigned to attend to her. The same doctor was instructed to attend to the fifty-two-year-old woman patient after that.

At 3.00 a.m., the woman said angrily, "When is my turn? I am in pain. I have a colonoscopy in the morning." She was informed that she would be attended to after the doctor had seen the other patient who was in severe pain.

"But I am also in pain," she complained.

"Just go back to your bed; you will be attended to shortly," I instructed.

At 3.30 a.m., Dr Kang sought my help. "She insists that we give her an anti-rabies vaccination." Dr Kang said.

"Why?" I asked.

"She says dog bites are dangerous, and she can die from rabies," Dr Kang answered. "She says that she is willing to pay for the vaccination."

I sighed. This was definitely not an uneventful shift. Everyone is a doctor.

Singapore is rabies-free. There is no danger of dying from rabies if bitten by a dog in Singapore.

"I know you," the quirky woman said as she wagged her right index finger at me. "I don't like your attitude. I thought you were a staff nurse, but now I know you are a professor. You know, I am also a professor at university. I teach business management, you know. I know I must have a rabies vaccination."

"Who told you that you must have rabies vaccination?" I asked, mentally shaking my head at whoever that was.

"I googled," she answered.

It was difficult to keep a straight face and not laugh out loud.

"As a professor in university, you should know that what is in Google is not 100 per cent correct," I said.

I had to put in an effort to speak in a calm voice, although I could feel my face flushing. I informed the patient and her sister, "Just because a patient wants a particular medication, that doesn't mean we doctors will give it to them. We have to weigh the benefits and the risks."

"So what are the risks?" the woman asked petulantly.

"You can die from anaphylactic shock," I informed her.

She stared at me. I had a suspicion that she did not understand what an anaphylactic shock was, but she did not want to show her ignorance. There was dead silence in the cubicle. I looked at her right thumb, where she had been bitten by her neighbour's dog. The wound was superficial – a one-centimetre abrasion. I was not inclined to continue with the conversation. I left with instructions to Dr Kang to proceed with wound care.

At 3.45 a.m., Dr Kang informed me that the patient had accepted that she did not require rabies vaccination.

At 4.00 a.m., this woman waved good-bye to me as she left the emergency department, saying, "Thank you, Doctor."

She beamed at all who she walked passed.

I was bemused for a few seconds at this 180-degree turnaround. I had expected to return to her cubicle to reinforce what I had said earlier. Dr Kang did well.

The shift continued.

The Conversation Meanders

The Big-hearted Patient

"Jasmine, was looking for you the other night," I informed one of my companions.

"Is she okay?" she asked in a worried tone.

"She's fine," I reassured her. "Just wanted to say hello."

Jasmine was a patient with multiple medical conditions.

One Nurses' Day, she gave each nurse in ED a rose. We said thank you but asked her to save her money; she told us that it was a blessing to be able to give.

On one Christmas Eve, she appeared in our ED pantry. I remember that moment, it was 7.30 p.m., raining cats and dogs, and there was Jasmine huffing in as she pushed a wheelchair with a roasted turkey, all the trimmings, a leg of ham, and mince pies, to celebrate Christmas with us.

Continuing the Conversation

"She really said thanks?" one of my coffee-drinking companions asked with widened eyes.

I nodded, and all of us burst out laughing.

Our second cups of coffee were half-empty by then.

I shared the following just before we called it a day.

It was about 7.00 p.m., on an afternoon shift when I walked into the resuscitation area to look for my registrar. I found him holding an ultrasound probe against a decidedly large abdomen. The patient was a slim twenty-four-year old woman lying quietly on a trolley. She had appeared at the entrance of the ED doubled up and barely able to walk without assistance. The staff had quickly put her on a trolley and pushed her into the resuscitation area; she had been screaming, presumably in pain. She had been asked by my registrar whether she was pregnant and had loudly denied that she was.

The screen of the ultrasound machine showed a small heart beating within the patient's lower abdomen.

My coffee-drinking companions shook their heads and asked in incredulous tones, "She really didn't know? Really?"

We may believe that a fourteen-year-old girl would not realize that she was pregnant, but not a twenty-four-year-old woman.

This fourteen-year-old girl was brought in to ED by her unsuspecting mother when she complained of abdominal pain. She was rather plump, which may have camouflaged her condition. She arrived on time because the baby was crowning.

It was an easy delivery. The grandmother did not look too upset when she found herself with a new grandson.

My coffee-drinking companions were chuckling away after I recounted these incidents. We were rather relaxed by then; our adrenaline levels had come down.

* * *

CHAPTER 10

OLD FOLKS IN THE ED

Let me take you through my encounters with old folks in the various parts of the world where I have worked. These are a few of their stories.

It was a winter shift in 1990 when I first felt that I was working in a different culture. I had been working in Edinburgh for four months, since August that year. I disliked winter, and that particular winter day was not a pretty one. There was snow, but there was also ice. The sidewalks were slick with black ice.

We were attending to a few elderly patients in the A&E department (this was what the Royal Infirmary of Edinburgh, Emergency Department, was called in those days). One of our elderly patients had been walking along the pavement outside her house when she had slipped and fallen on her right outstretched hand. She had injured her right wrist, which was shaped like a dinner fork. Her right wrist required manipulation to reduce the fracture and to re-align the bones.

It was about 2030 hours, and it had begun to snow again. The plan was to keep her overnight for observation and then send her home after arranging for her home support with the social services. She stayed alone with two cats. This eighty-year-old Scottish patient, requested that I called her daughter, who lived on the outskirts of Edinburgh. She wanted me to inform her daughter about her accident and have her come to the A&E

department. I receive this request regularly from patients in Singapore so I did not think that it was an unusual one. I went to the counter to make the phone call; the patient had given me her daughter's number.

"Good evening, may I speak to Mrs Beth MacDonald?" I said.

It was the daughter who picked up the phone. I introduced myself as a doctor calling from casualty (another old term used to identify the ED) and quickly reassured her that her mother was fine except for a wrist fracture. I conveyed her mother's request for her to see her in the A&E department. There was a significant silence on the other end of the line. I thought the line had been cut off.

"Mrs MacDonald?" I inquired.

"Please tell her we will visit her tomorrow, in the morning. Thank you," the daughter said before putting down the phone politely.

I was taken aback and a little stunned. It was not the usual response that I have received for such a request.

The patient's face fell when I conveyed her daughter's message. She did not say another word about the matter throughout the rest of her stay with us. I do not know whether her daughter came the next day because it was my day off.

It was probably the sensible thing for the daughter not to drive over to the A&E department that night. I discovered later that it would have taken an hour for the daughter to travel from her home to the A&E department – and likely longer because of the snow.

The usual responses I would get in Singapore would have been questions about the patient's well-being, how it happened, and whether they could talk to their parent over the phone. Then the next thing would be the arrival of a bevy of children or grandchildren in the ED. All the arrivals would be clamouring to see the patient. It would be a matter of too many asking for information at different times, and occasionally of many opinions. We coped because we were happy that our patients had people who cared and were concerned about them.

However, in our working environment it can deteriorate to an unhealthy reversal of roles between a parent and a child. We had a seventy-year-old Chinese gentleman in our resuscitation area. He had presented with chest pain, and we had made a diagnosis of an acute myocardial infarct. We wanted to transfer him to the invasive cardiac laboratory as soon as possible to balloon the blocked arteries of his heart.

Everybody and everything was in place. The only thing missing was his consent. We explained to him that he had had a heart attack and that time was of the essence. The sooner we carried out the ballooning for him, the more heart muscles we could save. He refused to make a decision. He wanted his eldest son to make the decision. Everybody and everything came to a standstill.

We called his son, but there was no answer because he was driving. It was another fifteen minutes before he arrived. The son made the decision for the procedure after we spent ten minutes explaining to him the risks and benefits of the ballooning and his father's condition. We had explained all this to the patient earlier. We finally rushed the patient to the invasive cardiac laboratory.

This contrasted with a similar situation I encountered when I was in the ED of a Connecticut hospital in the United States.

She was an eighty-year-old Caucasian lady with a full head of white hair. She had chest infection with heart failure. She was stable but required hospitalization. I remember that she was on a patient trolley, sitting up and wearing a hospital gown. Her son was standing on her right and facing her, and behind him was a brown-haired Caucasian lady carrying a two-year-old boy who was quietly playing with a toy car. My supervisor was standing on the patient's left, explaining to her why she required hospitalization. I stood at the foot of the trolley at an angle where I could see the patient, her son, her daughter-in-law (I assumed), and my supervisor against a backdrop of glorious autumn colours.

This was one week after 9/11. The emergency department was very quiet those days, and the staff was very sombre.

My supervisor looked at the patient, waiting for her to agree to the plans. She had been nodding and following the explanation being given. She then turned to look at her son, and before she could say anything, her son said, "Don't look at me. It's your life — you decide."

There was silence. My supervisor, her son, and her daughter-in-law looked at the patient expectantly. My supervisor did not look surprised or shocked. If anyone had looked at me, they would have seen my jaw drop momentarily before putting on a bland, professional expression again.

To this day, I still receive requests, instructions, and orders from children whose parents have underlying cancers not to reveal to their parents the diagnosis, because their parents have not been told previously. Sometimes I find that these children are deluding themselves: it is difficult to accept that their parents are unaware that there is something seriously wrong with them after being subjected to a barrage of tests and consuming new medications.

Most times I will work with these children to avoid opening Pandora's box. The contact time at the ED is short and limited; it would not be kind to the patient to reveal life-changing information and yet have no time to support them through the revelation – assuming they really are ignorant.

A quarter of a century after my experience in Edinburgh, I realized that our norms in Singapore have shifted towards what I saw in the West.

Is this for the better or for the worse?

CHAPTER 11

THE FRUSTRATED
'BIG BROTHER'

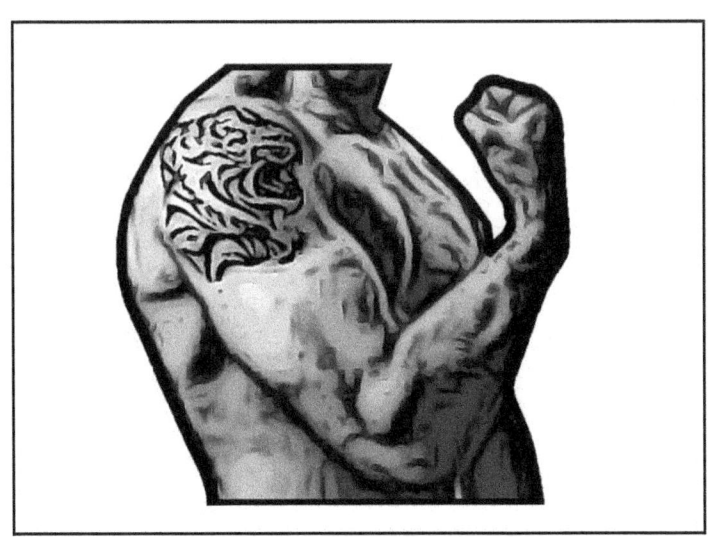

"Achek, lai, jeh lai kong." I said in Hokkien to a gentleman patient, in his late-thirties who was banging the top of our centre counter.

Literally translated, I had said, "Sir, come, sit down, talk." Or to put it elegantly, "Sir, come, let's sit down and talk about it."

The centre counter was where our counter clerks receive phone calls, discharge patients and sort out the various administrative tasks involved in the management of an ED patient. There were two counter staff (petite young girls) who were cowering as far away as they could from the patient. He had been cursing in Hokkien, which is a colourful dialect to use for such purposes.

The moment I had spoken, he became still. When I indicated that he should follow me, he did so quietly. We went to the consultation room just left of the counter, and I invited him to take a seat as I sat down myself.

This gentleman, Mr G, was well built, and his bulk was due to muscles rather than fat. He waited for me to speak. He wore a short-sleeve shirt and Bermuda pants. I noted the tattoos over both his arms, legs, and exposed parts of his chest and back were well drawn and had colours. However, the gentleman did not wear the customary thick gold necklace or bracelet.

I looked him in the eye and said in Hokkien, "Wu simi dai zi, hou, hou kong." Generally translated: "Any issue, say nicely."

He then told me in a calm tone that he had seen a doctor more than an hour ago, and he had been waiting to collect his prescription to get his medication. He had asked the counter staff a few times, but no one had helped him resolve the problem. Each time he asked, they informed him that they had not received it. I nodded as he related his experience. I understood why he had been frustrated and had lost his temper. This happened during a slow afternoon shift with not many patients around, which was fortunate. Otherwise, it would have been a threatening sight for most.

After hearing him out, I asked him to wait for me in the consultation room while I solved the problem. I did not draw the curtains but kept the room open so that he could watch the activities going on at the counter. The two girls at the counter kept themselves busy. I asked them what had happened. They said that they had not received any of the patient's paperwork. I went to locate the doctor who had attended to him and found him in the resuscitation room. He had been occupied in the last hour managing a badly injured motorcyclist. He had left the prescription and discharge instructions in his consultation room and had forgotten to inform anyone about Mr G. I sighed and apologized profusely to Mr G.

The team had taken my encounter with Mr G in a matter-of-fact manner, although I thought it was rather unusual that this well-tattooed gentleman had calmed down when I spoke to him. I had put it down to me having plenty of white hair and looking like a senior doctor.

When I related this incident to a retired police officer, Mr Low, he burst out laughing and could not stop laughing for a few minutes.

First, he informed me that the patient must be quite a senior figure in his gang because he had coloured tattoos. Next, he informed me that Mr G had shown me due respect not because of my white hair but because of my opening statement of "Achek, jeh lai kong". The phrase "jeh lai kong" (sit down and talk) is apparently the opening gambit to begin a discussion or negotiation for the people in Mr G's circle.

"He was probably surprised when you used that phrase and was not sure what your background or connections may be," Mr Low said, wiping tears from his eyes.

"You clinched it when you continued with 'wu simi dai zi, hou, hou kong'," he added.

I was dumbstruck for a while. A few minutes later, I was also rocking with laughter.

CHAPTER 12

WHEN EMERGENCY PHYSICIANS MAKE A CALL

In today's world, even teenagers can label a broken bone sticking out of a forearm as a fracture, hospital porters can recognize a patient in extremis, and astute nurses can pick out patients who can be rescued before they tip over. What special role, then, do doctors like emergency physicians have.

One of the emergency physicians' jobs is to treat when the cause is known. They decide the treatment options: go home with or without medications, discharge, hospitalize the patient, or (at the extreme) admit to an intensive care bed.

What happens when the diagnosis is not obvious despite the doctor combing through her memories of books studied and patients seen? It becomes even more of a challenge when the presentation of the patient's illness fits many diagnoses. Hopefully, the doctor has some ideas, and the investigations she orders will lead to the correct diagnosis.

A diagnosis is the label to an illness. The right diagnosis generally points to the right treatment. The wrong diagnosis can result in the wrong treatment for that patient, and sometimes it can cause harm.

The patients who shake the emergency physicians are those who have a list of probable diagnoses and have only a small (or no) window of opportunity for a confirmatory investigation before life-saving treatment has to be started. It becomes worse when the life-saving treatment itself can cause death if the diagnosis is wrong. In these instances, doctors' calls have to be correct.

A situation which can cause palpitations for emergency physicians are the close calls. This is when the team wipes the sweat off their brows, heaves sighs of relief, and says collectively, "Whew, that was close." This next case illustrates this point.

Mr Tan was a sixty-year-old Chinese gentleman who presented with right-sided upper and lower limb weakness. This had occurred three hours prior to his arrival at the emergency department. The stroke team was activated and their aim was to treat Mr Tan with a thrombolytic, a blood thinner, as soon as the CT scan of his brain was completed. The nurses were preparing the thrombolytic when the ED senior doctor walked into the resuscitation area. The nurses informed her of what they were doing. The neurologist who was in charge of the stroke team confirmed what the nurses had conveyed. The ED senior doctor glanced at the patient's face and decided to ultrasound the patient's abdomen. She found an aortic abdominal aneurysm which had dissected. The thrombolytic was not given.

The cause of the patient's stroke was dissection of the aorta up to a big branch, the internal carotid and down to the iliac artery, another big branch. If the thrombolytic had been given, the patient could have bled to death.

On the surface, this patient was a straight forward stroke case. When the ED senior doctor was asked what made her suspect that it was not, she thought hard about it. The only thing she could recall was that the patient's facial expression was not similar to that of other stroke patients' she had seen.

The whole team, including the stroke team, have palpitations whenever they recall this close call; the neurologist still breaks into a cold sweat today.

When a call goes wrong, the taste of defeat is a bitter one. It is part of the make-up of an emergency physician to be able to treat patients in extremis, even when a diagnosis has not been established. Most times the treatments allow them to snatch the patient from death jaws, identify the problem, and provide the solutions. However, there have been times when they have bought time for the patient but, because of the natural progression of the disease or because they were unable to identify the problem (or worse thought that they had and yet all treatments failed), they could not prevent the patient's demise.

Ms Sok was a forty-year-old Indonesian lady who was brought in by her relatives at about 1.00 a.m. She was running a fever and had a low blood pressure (80/50). She was in shock, and so two intravenous lines were inserted. Within three minutes of her being in the resuscitation area, 300 millilitres of fluids had been infused. The provisional diagnosis was septicaemic shock. Her oxygen saturation was 92 per cent on room air, but with 100 per cent oxygen delivered through a non-rebreather mask, her oxygen saturation climbed to 98 per cent. Bloods were taken for investigations, and her chest X-ray showed a patch on her right lung. The team had a cause for the shock: pneumonia. She was started on antibiotics, and more fluids were given. Her blood pressure climbed to 100/60. The patient looked better and more comfortable.

Thirty minutes later, her blood pressure dropped, more fluids were given, and an inotrope was started. She improved. An intensive care bed was obtained, and she was prepared for transfer. She had been in the resuscitation area for one hour, and as she was about to be transferred, her oxygen saturation dipped again, towards 90 per cent and her respiratory rate started to climb. The second electrocardiogram was done, and bloods were taken to exclude other causes. A decision was made to intubate the patient. Her blood pressure began to drop again. The inotropes were increased, and more fluids were given. The patient was successfully intubated, and her oxygen saturation climbed back to 100 per cent. Her blood pressure dropped further, and her heart rate, which had been hovering around 90 per minute, started to drop as well. Her blood pressure became unmeasurable as her heart rate dropped to 50 per minute and when the pulse was undetectable, chest compressions were started.

Ms Sok was pronounced dead ninety minutes after she had been brought into the resuscitation area. The team felt lousy. They had spoken to this patient and reassured her, and she had responded, and given them a thumbs-up; a relationship had been established. She was not a case to the team but a patient, a person. They could only pack her body now.

The calls that emergency physicians make are based on clinical reasoning (which depends on their intelligence), and sometimes just sheer luck. If it is sheer luck, is it the doctor's or the patient's?

> *"Clinical reasoning…. is the largely unseen skill that critically defines the performance of ED physicians." — Pat Croskerry*

CHAPTER 13

FAST FORWARD TEN YEARS....LOOKING BACK

Setting: A dim sum lunch get-together at the Imperial Treasure Teochew Restaurant at Ngee Ann City, Orchard Road.

Attendees: Six doctors who had worked in the same emergency department more than ten years ago.

Host: Ella, a retired emergency physician

Guests: Jon, full-time artist, part-time doctor

SH, civil servant, part-time emergency physician from Brunei

Gladys, full-time toxicologist, part-time emergency physician

Joyce, part-time emergency physician

Mic, full-time emergency physician

"Hi, bro, when did you arrive?" Jon asked as he clasped hands with SH.

"Have a seat, have a seat," I indicated to the seat next to Jon.

"Arrived late last night, and I thought I would be the first to arrive for lunch. I was expecting you to be late as usual," SH said with a chuckle as he looked at me.

I was well-known to be late for appointments.

"I was at Sandra's clinic earlier this morning. It is just across the road, and that's the reason why I'm not late. Here comes the rest," I said.

Gladys, Joyce and Mic walked in together looking surprised.

"We are not late," Mic said while looking at his phone. "In fact, we are slightly early."

There were greetings around the table. A few had not met in the last twelve months.

"I am happy that this restaurant is still around," SH said as he looked around.

"Lucky for Ella. It's one of her favourite makan places. The soups here are some of the best in town," Jon said.

"Let's order first. I don't know about the rest of you, but I am hungry." I said.

After we ordered our favourite dishes, we started to exchange news about the people we knew, our work, and the old ED.

"Gladys, congrats! I heard that you managed to get the regional toxicology network set up," Jon said.

"One of my goals achieved. Yay!" she said as she pumped her hand in the air.

"She has Brunei linked to the network as well," SH added in a proud tone.

"Most important, I have the pioneers in toxicology as advisors to the network," Gladys said with a grin.

"Speaking about pioneers, what happened to NBM?" SH asked.

"NBM? Who is NBM?" I asked in a puzzled tone.

"Jack's nickname was NBM - nil by mouth," Joyce enlightened me. Jack was an old physician they had worked with in their younger years.

"The nurses and medical officers nicknamed him NBM. His voice was so soft that they couldn't hear what he was saying. It was worse when he wore a surgical mask, and it was non-existent when he wore an N95 mask," Mic explained.

"The medical officers would ask him what to do, and then they'd ask again a second time when they couldn't catch what he said. The new ones would ask a third time, and Jack would click his tongue like a lizard and give them this annoyed look. They would still not understand what he said. After a while the nurses would take pity on them and decipher the instructions for them. The more experienced ones would turn to the nurses for the interpretation after Jack had spoken," Joyce added.

I looked in disbelief as the rest hooted in laughter.

"Remember, that group of medical students who were a bit rough around the edges?" Mic said. A few of us nodded. "Well, they invited me for a plate of char koay teow at the end of their posting," Mic continued.

"Why? Because you are a sam-seng like them?" Gladys teased him.

"Maybe," Mic answered as he shrugged his shoulders and smiled. "When I asked them which of the senior doctors they were afraid of, it wasn't you [pointing at me] or Kate. It was Jack."

The rest laughed as I put on a face of mock disappointment. "I thought medical students and officers were scared of me," I said with a pseudo-whine.

"This set of medical students said your bark is worse than your bite – no need to scared," Mic said in a mock comforting tone.

The rest could not keep straight faces anymore and burst out laughing. I joined in the laughter after a few seconds.

"They were not afraid of Kate, either," Mic continued.

"Why?" I asked as I wiped tears of laughter from my eyes. There were smiles all around the table. Everyone knew except me.

"They said they knew what Kate wanted because she talked and talked and was very loud. They couldn't fail to hear her. Just do what she says, no problem. Whereas with Jack, they couldn't hear what he said even when they asked again and again. They didn't know what he wanted. This scared them, not knowing," Mic explained.

"I am very disappointed. I thought I was the scariest," I said, laughing.

"You were pretty scary in my younger years. I remember how the nurses made sure they cleared the observation area before you did your afternoon rounds. They and the nursing officers would quickly transfer the admitted patients to their hospital beds so that there would be fewer patients during the round. If not, you would be asking why," Joyce said.

"Yes and I remember Viv [a previous colleague of theirs] disposing of her patients before 2.00 p.m. – either discharging or admitting them. She would get the medical officers to do the same. This was to avoid Ella hauling them to the observation room to ask why," Jon added.

The group burst into another round of laughter.

"You mean all of these were happening, and I didn't know about it?" I asked in astonishment.

"You didn't know all of these were happening?" Gladys teased me. "The nursing officers were very diligent in ensuring that the patients had their lunch and had gone to the toilet before your afternoon rounds."

The rest were in stitches. By now our group was receiving curious looks from the other diners. When the dish of roast duck arrived, the group moved on to talk about their recent travels.

"The dim sum here is as good as that in Hong Kong, and service is still as good as that ten years ago," SH commented.

"Only because it is Ella," Jon said.

The rest chuckled.

"I heard that the communication skills of the younger doctors are much better than their seniors since the hospital started teaching them how to read body language," I commented.

Gladys and Mic, who were still involved in the training of the younger doctors, nodded. "Much better," Gladys confirmed with a smile, "I remember one night shift when you pointed out a couple who were walking to a consultation room. You said that the woman was not the man's wife, maybe a girlfriend or mistress. You were right — it was his girlfriend."

"You asked me how I knew," I murmured.

"You said that the woman was 'too openly' affectionate, clinging to the man's arm, her body leaning against him, and walking as if she would fall without his support. It turned out the man was the patient," Gladys continued.

"Can the younger doctors spot this type of relationship nowadays?" I asked.

"They are pretty sharp and quite quick at spotting strange behaviour, as well as when people are lying," Mic said.

The conversation then revolved around the quality of doctors from the three local medical schools. Then we went back to talking about people we knew.

By the time dessert arrived, we had caught up with the happenings in each other's lives.

No Regret

It has been more than a quarter of a century since I first worked as a doctor in an emergency department. When I decided to specialize in emergency medicine, many of the concerns about — not being able to go private, not being my own boss, working shifts — did not loom in the equation. It was the excitement and the satisfaction of being pivotal in a patient's well-being that clinched the decision. There are no regrets at least for me and most of my friends in emergency medicine.

I asked a few of them this question, "What gave you the most satisfaction in your years as an emergency physician?" These were some of their answers.

"Helping people." — A Yip

"The chance to help people during an acute medical emergency." — CKC

"Satisfactory stabilization of ill patients and their management." — V Siu

"Life-saver." — Mic

"Making a difference." — Lee SW

Over the years, there have been many occasions when being present made the difference between life and death for a patient. The feeling of snatching the patient back when his heart stopped working and getting the heart to pump again is one of those bread and butter situations for us, emergency physicians. Each time this happens, I still have this euphoric feeling of having made a difference. Occasionally we fail to snatch a patient back.

Fortunately, for our mental health, most of our patients do get well. They improve with our treatments: a breathless asthmatic patient, when given a salbutamol nebulizer, breathes more easily; a patient with food poisoning improves, stops vomiting, and stops visiting the toilet; a patient with a laceration to the scalp has the wound sutured, and the bleeding stopped.

There are times when patients are at the end of their life span. We know it and we do our best to make the patient more comfortable.

To quote G Cham, an emergency physician for the last two decades,

"My happiest moments in the emergency department is when I see patients get well. If they don't get well, at least they feel more comfortable."

The positive interactions between patients, their family, their friends, and us are also a source of inspiration and satisfaction in our work. Emergency physician, LWY received a standby patient, and as the resuscitation room door was closing, she heard one of the relatives tell the rest, "It's okay, I've seen that doctor before. She [the patient] will be all right." LWY felt a sense of pride at being trusted, and she had confidence that she deserved that trust.

"The patients' and their relatives' appreciation of the care we offer." — V Siu

"There have been occasions when the relatives of patients who had passed away several months ago return to say thank you." — LWY

During our training years, we knew that working in the emergency department would be like working in a pressure cooker. We simply did not know how high the pressure would go, and fortunately none of us cracked. In this type of environment, one of the best gifts we received has been friends, whether doctors, nurses, porters, clerks, administrators, or students.

> *"Having made very good friends. The highly stressful and intense ED work helps bonding as we work as a team"* — *Joyce LSS*

> *"Able to work as a team, looking out for each other."* — *Ooi CK*

> *"Great team and colleagues, positive working environment."* — *Ang SH*

To this day, for me the emergency department is still a wonderful place in which to work. It is where I have the privilege to help, to save, and to make a difference in my patients' lives. It is also where I meet great people and make lifelong friends.

TILL WE MEET AGAIN....SOCIALLY

At a dinner, one senior doctor told another that he should carry my handphone number because they were of the age when they were likely to have to be at the emergency department as patients. I demurred and said, "Hopefully not".

Another said, "We have had to visit the place."

I murmured, "Hopefully not again for a long while." It is true that we never know when, even we emergency physicians will have to be in the ED as a patient, or bring a loved one in as a patient.

The ED is the only medical facility in the community that is open 24 hours, 7 days a week, 365 days a year to manage emergencies, especially the life-threatening ones. It is a peculiar place to most, but we hope this book has thrown a bit of light on the work that goes within. It is a special place for the people who work within it, though occasionally you may hear us grouse about the pressure, ungrateful patients or members of the public, the unfairness of the system, the stupidity of....But despite all these, we are still proud to declare that we work in the emergency department.

We who work in the emergency department are simple people, although we admit we are adrenaline junkies who thrive on the unpredictability of the place. We are social creatures. We still like people despite the horror stories we have encountered. The most important reason for our longevity in a less-than-benign environment ("Other than salary" says Joyce LSS) is a belief that we are making a difference.

Thank you for taking the time to know us better. We wish you the best of health.

Till we meet again…..*socially.*

GLOSSARY

12-lead electrocardiogram – tracing of the heart's electrical activity

adrenaline –a hormone that raises heart rate and blood pressure

AMA (against medical advice) form –this form is signed by a patient or his guardian when he declines admission or medical treatment when advised by a medical practitioner

aortic abdominal aneurysm – an enlargement in the aorta when it passes the abdomen; the aorta is the largest artery of the body

baluku – "hematoma" in the local lingo

cardiac event – heart event

COPD (chronic obstructive pulmonary disease) – a group of lung diseases that block airflow and make breathing difficult

crowning – the baby's head is visible

defibrillator – equipment used to pass an electric current to the heart elevated white cell count – usually indicates that the body is fighting an infection

ENT – ear, nose, throat

inotrope – a drug that alters the energy of muscular contractions of the heart or of the blood vessels

intravenous lines – putting tubes through veins

intubated – put a tube into the windpipe

myocardial infarct – heart attack

non-rebreather mask – used in medical emergencies to deliver higher levels of oxygen therapy

N95 mask – a mask that blocks 95 per cent of airborne particles

NSTEMI – non-ST elevated myocardial infarct; a type of heart attack

overdrive pacing – using a device to increase the heart rate in order to suppress abnormal heart rhythms

percutaneous catheterization (PCI) – ballooning of the arteries of the heart

per rectum – through the rectum

Pulse oximeter – a device that measures blood oxygen level

registrar – mid-level staff

resus – resuscitation

RTA – road traffic accident

sam-seng – gangster in the Hokkien dialect

septicaemic shock – occurs when an overwhelming infection leads to life-threatening low blood pressure

shock (in shock) – low blood pressure

shock – to pass an electric current through the heart using a defibrillator

standby case – a critically ill patient that the emergency ambulance team will transport and give prior notice to the emergency department of its arrival

systolic blood pressure – the pressure of blood against the artery walls when the heart beats; anything below 90 is generally considered abnormal

T&S – toilet and suture; to clean and stitch

ventricular tachycardia – rapid heart rate that occurs from abnormal heart activity

About the Author

Dr. Eillyne Seow has been an emergency physician for 25 years, a pioneer in Emergency Medicine in Singapore who has worked in several Emergency Departments in Singapore and other countries.

www.ingramcontent.com/pod-product-compliance
Lightning Source LLC
Chambersburg PA
CBHW051544170526
45165CB00002B/877